CW01159547

Sensual Woman

Sensual Woman

An exploration of the true meaning
and value of sensuality

Sarah Bartlett

LONDON HOUSE

First published in Great Britain in 1999 by
LONDON HOUSE
114 New Cavendish Street
London W1M 7FD

Copyright © 1999 by Sarah Bartlett
The right of Sarah Bartlett to be identified as author of
this work has been asserted by her in accordance with the
Copyright, Designs and Patents Act, 1988

This book is sold subject to the condition that it shall not,
by way of trade or otherwise, be lent, resold, hired out
or otherwise circulated without the publisher's prior
written consent in any form of binding or cover other
than that in which it is published and without a
similar condition including this condition being imposed
upon the subsequent purchaser.

A catalogue record for this book is available
from the British Library

ISBN 1 902809 00 9

Edited and designed by DAG Publications Ltd, London.
Printed and bound by Biddles Limited,
Guildford, Surrey.

Contents

Introduction, 7

Part 1, BODY SENSES, 11
1. The Sensual, 13
2. First Sense, 31
3. Bodily Diversions, 39

Part 2, BEAUTY SENSE, 51
4. Skin Deep, 53
5. 'I have perfumed my bed', 71
6. Girdles of Desire, 83

Part 3, PANDORA'S JAR – SENSUAL ILLUSIONS, 99
7. The Rustled Sheet, 101
8. The Dark Feminine, 113
9. 'Fair is foul and foul is fair', 127

Part 4, SENSUAL GRACE OR DISGRACE?, 141
10. 'Mary, Mary, quite contrary', 143
11. Feeding Frenzy, 159

Part 5, BEYOND THE LOOKING GLASS – THE VALUE, 173
12. Romancing the Senses, 177
13. Sensual Eroticism, 193
14. Why Men Need Seducing, 213

Part 6, PLEASURE HUNT, 229
15. Femmes Fatales, 231
16. Glamour, 249
17. Sensually Yours, 265

Last Words, 277
Bibliography, 281
Index, 285

Acknowledgements

Thanks
to Rod for being the catalyst;
to Larry for being a sensual man;
to Darby Costello and Maggi Pagano
for being inspiring friends

Introduction

Sensual: of the senses, carnal, worldly

This book is not going to tell you how to be sensual. It is not a do-it-yourself, self-help sex manual on how to give pleasure to your partner, or how to get in touch with your body for sheer enjoyment. If it were that kind of book we might be missing the point, which is that we are all already sensual. It is a quality we already possess just by virtue of being human. And so the point of this book is to give true meaning to sensuality and honour it, before it's too late.

The sensual is a quality which permeates nature. In every living thing the senses are the primitive instincts, the *prima materia* or base substance from which we set out on our fundamental quest to be in relationship with our perceived reality. This sensual quality is one of the energies that works behind and through life, and however far away from our instinctual nature we believe ourselves to be, the sensual sends ripples through our words, ideas, images and feelings. *Sensual Woman* reveals how we have arrived at labelling woman as an exclusive embodiment of the sensual, and how men have suppressed or denied their own sensual energy because of historical, cultural changes and the urge towards civilization.

For thousands of years, ever since mankind separated itself from nature, both men and women have participated and colluded in the notion that being sensual is an exclusively female quality. This book explores the the many illusions surrounding what sensual is and isn't, and how the image and symbol of masculinity has not happily corresponded to the sensual. It also reveals how we have bought into this myth which, like any myth, holds the key to a profound and hidden truth.

Sensuality is genderless, but in Western civilization the sensual is seen as a feminine attribute. Woman has become the ultimate symbol of sensuality and mankind's fate. She is the diva of earthly pleasure in the grand opera of life. She is both life-giver and breath-taker. Paradoxically, women are perceived as embodying a quality which has either been enriching, erotic and cherished, or repressed and distorted by many cultures.

This book looks at the historical evidence and the changes in Western and Eastern perception that may have occluded the truth about the sensual world. It explores those women who have been considered particularly sexually sensual, from Cleopatra and Helen of Troy to Marilyn Monroe, *femmes fatales* such as Mata Hari, seductresses and temptresses real and imagined, including La Reine Margot, Christine Keeler, Lilith and Eve. It looks at how the words and the ideas evoked by the senses have become synonymous with sexuality both historically and psychologically.

For thousands of years, women have been perceived as the prime symbols of nature's evil, the temptresses or succubi of men's precious life-force. This book reveals how sensuality became fused with sexuality, and femaleness was idealized or debased. The ideal woman became like a Barbie doll, clean, bendable, obliging, and vulva-less. Ruthlessly compartmentalized, ordinary woman was either virgin and mother lacking in sensuality, or whore and seductress, the archetypal siren. This is still a nagging unconscious dilemma for Western men.

The wild, free, civilized, wise woman is undergoing her own renaissance, but are we in danger of devaluing and denying sensuality? No longer in need of breadwinners, women are still expected to be sexual enchantresses at night, while they choose to be workaholics and mothers by day. Gradually, we are losing touch with the enchantment of the night at the expense of daytime materialism. Our intimate relationships are chained with dangerous illusions and high expectations, and we see the sensual world sublimated into pornography or obsessional eroticism. This book suggests that the sensual is an archetypal quality, an experience which not only filters

INTRODUCTION

through the human machine, but is an energy which also pervades the essence of the soul and the universal matrix. We need to know what it is we are in danger of losing, and why it is of value to us.

The price we may have to pay for dishonouring our sensuality might be higher than we think. To nurture only the power of progress, arrogance and materialism is to speak only one language. It denies all others, cuts us off from our bodies, and subsequently leaves us soulless. This book intends to show how woman has been cast as the diva of the senses, yet paradoxically, it is that very sensuality, when fused with the power of female sexuality, which has been humanity's redeeming force. It also explores why men find it difficult to integrate the notion and the experience of being sensual, and why women may be gradually denying their own feminine sense. It explores the true meaning of sensual enchantment and why it is of value in a disenchanted world. Losing touch with the one earthly delight which is our fundamental human need means love may become just a game played on a mobile phone.

PART 1
Body Senses

CHAPTER ONE
The Sensual

*'All things are connected like the blood which unites us all.
Man did not weave the web of life, he is merely a strand in it.
Whatever he does to the web, he does to himself'*[1]

The Sensual and Woman

This book has a deliberately misleading title. It evokes an immediate image in one's mind that woman and the sensual are synonymous. Isn't it difficult to imagine the word 'sensual' without thinking of women, or to imagine 'woman' without thinking of words like sensual, seductive, alluring?

For thousands of years men and women have designed their relationships according to a collective perception of man as potent and dominant, and woman as sensually alluring. In one sense the sensual seems to be more easily embodied in a woman's body, created as the female is for receptivity, the preservation and carrying of life. If women have for thousands of years been solely identified as sensual, and men not, what is it that caused us to become so fixated about sensuality being exclusive to women?

Firstly we must attempt to look at the sensual as separate from femaleness. Then, when our minds are unclouded by gender perceptions, we may be able to see how a woman and man are the most evocative symbols of the 'feminine' and 'masculine', or what the mystical East calls yin and yang.

'Symbol' is derived from a Greek word meaning something thrown together. We infuse ideas, objects, and feelings with symbols. For example, emotion is symbolized by water, or aggression by fire, but equally water itself is symbolized by feeling, and fire by dynamism or potency. Symbols are the underlying, overlaying, interplaying energies of the invisible through the visible.

Symbols infuse meaning into our sense of being, they are thrown together like man and woman, a continuum of binding and creating the symbolic in itself. For example, water, woman, art, night, and the dark are all symbolic of the 'feminine' or yin-sense, and fire, man, logic, power, and progress are all symbolic of the masculine, or yang-sense. Femaleness is also a feminine symbol, as is maleness a masculine symbol.

The sensual means 'of the senses', and is a quality which permeates everything, it is the stuff of which we are all made, another of the base energies and instincts of humanity we all share in different proportions, or are allotted while alive on this earth. Some of us are born with a deeper sense of pride than others, others are born more humble. Some are born aware of their bodily senses or their instinctive nature, others are less aware and more disposed towards the abstract or mind-sense. Some of us are acutely sensitive to other people's moods and feelings, while others have a more pronounced sense of ego or strong emotional boundaries. The qualities or energetic senses that we do not express or make conscious, we relegate to the unconscious, and they are then only discovered through our relationship with the world. We meet ourselves and our hidden energies in those we are attracted to or are repelled by in life. These unconscious qualities are best known as our shadows. What we do with our shadows and outcasts is discussed later in the book.

Making Sense

'Sense' comes from the Latin *sentire*, to feel, and also the Greek *ousia*, the root of the word 'essence', 'that which is one own's property'. Yet we have phrases such as 'come to one's senses', meaning to regain consciousness or get back to rational thought, also 'it must make sense'. We know of five bodily senses: sight, sound, smell, taste and touch. Then we add awareness, the sense of knowing, and another we have defined as intuition, a mysterious sense that cannot be measured or defined by science. Again, intuition is given a 'feminine' label. We have coined this mysterious sense of knowing as women's intuition, not men's. But in itself, intuition is simply a sense awareness, it is neither male nor female.

To make sense, meaning to make rational, to see the explanation of something, is also used to bring something into consciousness – an apparently masculine quality. Is there not some contradiction here about our senses, that they are both carnal, experiential, feeling and pleasurable, yet rational, conscious and explainable? And why is it that the rational senses are endowed with a masculine quality, and the feeling senses a feminine quality? Each of these authentic sense-qualitites are archetypes. James Hillman describes archetypes as 'the deepest patterns of psychic functioning, the roots of the soul'. He imagines them as 'fundamental metaphors', holding 'whole worlds together'.[2] Archetypes are neither masculine nor feminine, but they can be symbolized as either. They are the fundamental experiences that permeate every living thing. Constantly in dialogue, perpetually symbolic or 'thrown together', the diversity of the sensual binds us to others as it loosens us from subjective illusion.

Bodily Senses

The bodily senses are those with which we can identify most objectively. Touch is the sense we associate mostly with sexuality. From the moment we are nestled in the womb, and from the moment we are born we are induced into the world of touch. To touch and reach out to mother means we can survive – our mother's touch is our lifeline. We immediately respond to her caress and the touch of her skin against us. Sight, sound, taste and smell are also primitive sense states. We need to develop them as we mature, but touch is contact with the world: it is our first relationship.

Primeval messages are generated between two bodies just as when you generate your own personal sense of touching a table, stroking a cat or handling a silk dress. Sight, sound, taste and smell are not shared between two people in quite the same way as touch. Lovers may see each other, gaze into one another's eyes, they may listen to the voice of the other down the phone or across a restaurant table, they may taste each other's kisses or mouths, and smell the perfume and fragrance unique to each. But these senses evoke the present moment and memory, rather than the future moment and its possibility. When two people touch, the future is evoked, the other senses are generated, and there are more feelings, thoughts, images and moods conveyed than with any other sense.

Forgotten Senses

We tend to forget or repress the other senses. In other words, our perception of the world around us, our 'egoic' sense, the sense of fear, or the sense of love, are often taken for granted and historically have been regarded as having little to do with our sensual world. We also forget our sixth sense, that of intuition or divining, and the other sense of feeling, of being aware of the outer senses and being aware of feeling 'feeling'. Again, we don't often realise that when we are conscious of something we are also sensing the awareness of knowing. These senses are vital if we are to evoke the sensual and give it value in all areas of life, especially in our sexual relationships. Intimacy binds, and it also can destroy, for the sensual world means we experience both pleasure and pain.

Taking pleasure in the senses, which we call 'sensuality' demands of us a response. It also demands that we are in relationship to the world, the most fundamental aspect of living. Without a vision there would be no seer, without a sound there would be no one to hear, without a soft skin there would be no one touching. But equally, without a voice there would be no sound, without a tongue to taste there would be no sweetness or bitterness. There has to be a relationship between the self and the environment whether it is another person or a flower, for sensuality to even manifest itself.

Sensual Love

Sensual love is our delight in our earthy, feeling nature. Taking pleasure in the senses can be likened to the fusion of bodily pleasure with the awareness of love. Sensuality is not just to give and take sexual pleasure. People can be sensual without being sexual, or they can be both.

Another quality that evokes the sensual is the sense of mystery. It is feeling of the emotive kind, of the mysterious intertwining of love, desire and belief. It is the interplay of profane and sacred energy, the mystical union of yin and yang. Taking pleasure in the sexual senses is often the most powerful and creative way by which we can express this natural mystery. Our sexual relationships are often the only place we encounter these sensual realms as adults, and as children we encounter our first sensual confusion through our perception of our parents and our early relationship to them and others.

Being sensual means participating in the experience of being part of nature, however big we think nature is. A fly on the wall, the first kiss, waves crashing on a beach, are all what we call 'reality', meaning we must engage our senses. To be sensual is to be in relationship with the world. It means being at one with the tangible body of oneself, and therefore the body of mankind.[3]

Female and Male, Feminine and Masculine
The Same Difference

Men and women are similar in soul, feeling and mind. Biologically however, we display visible differences, our bodies created for different functions to maintain and prolong the species. From this difference, the collective assumptions of gender we are weaned into from birth emphasize the already basic animal distinction. Historically, this fundamental biological difference has been observed, examined and rationalized. If we are different in body, then the question arose that we must also be different in intellect, ability, emotion and soul. We are animals with consciousness, and as consciousness changed so did our sense of him and her, this and that, me and you.

The human species, like any other life-force, has an instinct to survive. At the most profound level we must be captives and captors, man must hunt woman, and woman must draw him in. This is our basic animal drive, and it is has led to much confusion about our motivation in life. Freud believed we were motivated by biological needs, Jung by the urge to wholeness, and Adler by the will to power. It seems likely that our underlying motivations are a synthesis of all three, a three-fold arrangement imagined by the psyche.

The basis of our enslavement to one another is the biological fact that women have secret entrances to their wombs which men must discover and respond to in a potent way. They must rise to the scent, then hunt for the eggs they must seed. But even when we are not playing the mating game in the biological sense, we still hunt one another. We are bound to one another because we are the same difference. We must capture and be captivating, we must allure and be lured, we must be beguiled and be beguiling. The problem mankind has with this is that immediately it seizes on the polarity of opposites, it names and classifies, splits and separates, particularly in our Western heritage. We cannot see that the same difference is in fact the different sameness. The only boundaries between men and women are the ones we have created in our quest for consciousness. Woman-ness exaggerates the energies of anything perceived as feminine, as male-ness exaggerates anything perceived as masculine.

Many men have suppressed their 'feminine' nature, because they simply had to shut off their feeling zone to get anywhere. We may ask why they need to 'get anywhere'? Which brings us back to the dilemma of consciousness. Without consciousness you cannot ask such a question. We have conditioned ourselves to follow the man-woman difference for over twenty thousand years. Dug into our biological database, our genetic codes and our inherited consciousness is an image of woman as vulva/womb and man as phallus. The vassalage and bondage of man to woman and woman to man is archaic, threaded with distortions, transcendences, ripples of love and waves of confrontation. The dance is played out on the world stage, and the difference magnifies, expands, eroticizes and moves us ever on. Wilfully, we refuse to see the parity between us – the disparity is always more intriguing. It is easier to deny that men and women could ever be the same. No, she is a woman and wily; no, he is a man and destructive. Yet woman can be destructive and man can be wily. We have put words into one column headed 'male', and other words in a column headed 'woman'. We have imbued language with gender. We are gripped by polarizing everything into light and dark, black and white, and male and female. Light is masculine, solar, potent; darkness is feminine, lunar, hidden.

Both men and women are a 'coincidence of opposites' themselves. Both harbour and opportune, both desire and defend, both have their own feminine and masculine images and qualities. Yin and yang are an interplay of energy in every person, whether man or woman, as in the cosmos.

Male and female are the most genuine examples of the fundamental biological difference, and accordingly the most obvious examples of the dualistic perception of the world. It may be the world-soul set up the animal kingdom as primarily male and female, rather than parthenogenetic, as a more random and chaotic generator of the species, to alarm us, to keep shocking us into seeing the very difference which like any polarity merges into union as One. Many philosophers, such as the Taoists, and the early Greeks like Heraclitus, believed that the true nature of the cosmos is maintained by the tension of opposites. Through this dynamic tension permeates the 'ever-living fire', or life-force. Biologically, man and woman are prime examples of the tension of opposites, but we seem to have forgotten that the life-force and the psyche, or inexpressible 'incognita' of the world-soul, permeates men and women equally. It is only in their unexpected embrace that the polarity is magnified. The sensual sends ripples to those who will run their fingers through 'her' still water. The sensual, like the unknown, may have been classified as 'feminine', but 'she' shares the same fire, air, earth and water in every man as 'she' does in every woman. We need to look at history or her-story, to discover why so many men have become split from the sensual, and how women have become a symbolic exaggeration of all that is deemed feminine, especially sensuality.

Nocturnal Fruits

Once upon a time, mankind's consciousness was so fused with the environment that we were at one with our senses and with nature. There was no sense of a separate self. We were not men and women, we were just part of the landscape and of the body of the earth. There is an ancient Aboriginal creation myth, albeit retold over many thousands of years, which suggests the androgynous nature of the first people.

> 'These were known as the Djanggawul, a brother and two sisters. The two sisters had huge genitalia, a combination of both male and female organs, which they dragged across the desert, causing great trails in the sand. But one day brother-Djanggawul, with only one penis, cut off his sisters' huge distorted phalluses, leaving only gaping wounds, swollen

open flesh that would forever be sacred. This was where Djanggawul-brother would now place his own penis, and thus become the superior creator of the Aboriginal race.'

In a way the Djanggawul story is symbolic of our initial split from nature. The flowering of consciousness, the mind noticing it could differentiate between me and it, and then he and she. His body was different from her body, but why was an still unknown word. Heaven and earth became separate too as mankind began to look up and notice the patterns in the sky that kept repeating themselves. They began to make sense of these patterns, and use them, work with the daylight to hunt and gather, and sleep at night, or take their first nocturnal pleasure.

Like Djanggawul brother, prehistoric mankind may have separated from the power of nature in an attempt to contain and control it. Women were child-bearers, child-minders, healers, carers and protectors. It is likely that they had little time for dreaming, visions and abstraction. Birth and death were not so much beginnings and endings as key images in life. Mankind had no choice but to cut off from nature in order to evolve.

Much later there was a time when neolithic humans found they could paint images on cave walls, and mankind began to enjoy the notion of me, I, 'my' possessions. It was all about becoming civilized. Nature was still powerful. Animals, beasts, plants, storms and the wind became symbols of the mystery of the separate self. Now nature and the divine could be fashioned into art. The divine could be found in nature, and nature became sacred, animals were sacred, and man and woman was sacred. But women seemed mysterious. They bled and they carried children, and somewhere in these distant dark ages, men began to realize that they had something to do with fathering. Once the seed of knowledge was sown, men began to master themselves, and the world around them, including women. Becoming aware of their need to hunt for food and women's eggs meant that the passive, yielding, feeling side of their nature was less important. Ironically, man almost castrated himself from his own closeness to nature, just as the Djanggawul cut off his sisters' genitalia, and along with it, his androgynous connection.

Symbols and Nature

Mankind has always had a love affair with gods and goddesses. These are our immortal interlopers, our clandestine go-betweens from the soul of the earth or the archetypes that permeate the human imagination. The current cauldron of 'goddesses' we are boiling up on our space-age hobs to give the 'feminine' a new meaning in life, is simply a projection and expression of our age and culture. If you have a god or goddess to live by then life is easier to manage. Nature can then become the fantasy. It can be the stage for the drama of life to take place. The props can be changed, the scenery rearranged by the divine director, and we, the actors, can come and go, as long as we play our parts. Gods and goddesses free us from responsibility, and from the disturbing, indiscriminate sense of nature.

But this underworld and overworld and inner world that we nearly manage to experience through dreams, fantasies and imagination, bursts through into our egoworld when the gods and goddesses come clamouring to enrich or deplete our lives. They take us nearer to the truth behind the illusions, they remind us that there is an inexpressible mystery, a moving, organic inseparable wholeness that weaves itself into and out of human life, just as it weaves into and out of all creatures, and the cosmos in its entirety. As Plato wrote: 'The region beyond the sky no poet in the past has hymned as it deserves nor will any in the future, but it is like this . . . the region belongs to being as it really is – without colour or shape, untouchable, perceptible only to the soul's charioteer . . .'[4]

The symbols of the invisible are unspoken languages which we put into words. But the language of the soul, or mana, or ch'i or prajna, comes to us through the energies or archetypal experiences that also appear as the gods and goddesses of imagination. Or perhaps, like Vishnu, it is the gods who imagine us.

Romancing the Past

Ironically, we have cut ourselves off from our basic nature through the very sense we love most, consciousness. Jung believed we were deeply connected to the ancient past and nature through the collective unconscious. Briefly, the further back in time we go, neolithic and beyond, the more likely we are to meet the most amoral instinctual energies that are swimming around in our own unconscious. Gods and goddesses are metaphors for these basic primitive forces, as the energies are metaphors for the gods. As a race we are trapped in the forgotten memory of the first ancient man and woman as they first copulate, as they first discover feeling, as they first discover children, as they first discover sexual pleasure, as they first discover they are separate from one another. This ancient psychological shock lives within our collective unconscious psyche, and we really don't have any chance of shifting it. Jung's philosophy may well be a fancy of a man who was bewitched by the psychology of his own nature and time, but it seems to ring true. Perhaps we are just beginning to realize that the distant past is a very special place, as long as we can look back without anger and remember we are also romanticizing and idealizing the ancient world, coloured by current cultural needs and individual perceptions.

Woman and Time

Woman's body is a sacred and secret place. It is hidden, secret and veiled in possibilities. It is pagan, demanding and cyclical. It has come to symbolize nature's force and nature's power.

It seems women have throughout history symbolized an ancient primeval quality, which has at times led her into the deepest water. There have been many arguments for restoring the Mother Goddess, or a golden age when women and peace were synonymous. Whether this can ever be proved, or is just another illusion of our current psychological neurosis, still fascinates and confounds us. Some argue that there were Great Goddess cults prevalent throughout prehistory, and real women had power and there was peace. Others staunchly refuse to agree. Intuition is not enough: there is no evidence, there are no documents, there is no reason.

Both viewpoints are probably right for this moment in time. This is a time when ambiguity, paradox and contradiction, the illusion of reality and the search for meaning in life has reached another renaissance. It is timely to look back, because that is the nature of our time. The nature of the sensual may seem a small molehill on the bigger landscape, but it is part of the same defence, or the same attack.

This is the stuff of the dark sensual world, the one we prefer not to venture too close to. But the sensual world can be pure pleasure. It is both the darkest night, and the sun rising across the misty field of molehills. The sensual is the one diversity in form, in experience, in being, in pleasure that keeps us in touch with the 'shape-shifting' qualities of nature and its eternal dance with culture.

Body Art

The secret sexual body of woman holds caverns, hills and a landscape harsh yet tender. The ambivalence of woman's body is attuned to the rhythm and natural forces of creation. She may contain a child if she so desires, and she can enjoy her body and take pleasure alone if she so choses. The male body can do this also, but biologically their bodies are for seeding, not fruiting. Humanity has struggled with the physical appearance of the female form for thousands of years. It seems from early archaeological evidence there was an ancient time when pleasure, particularly sexual pleasure, was sacred and celebrated.

Fertility and sexuality don't necessarily walk the same path. There was nothing wrong in the joy of sexual pleasure until someone decided there was. Woman is unique among animals in that she alone can take great sexual and sensual pleasure in her body, without involving her reproductive organs. Biology lessons may send girls into giggles at school. Women may blush as red as their vulvas at the mention of the clitoris, but the female body is extraordinary, and it seems it may have got women into an awful lot of trouble in the distant past.

Women have this secret place, a womb. It seems tomb-like, suffocating, pulsating with life, then empty of life. A woman has an instinct to preserve this place for the growth of life. It may be imagined as an entrance, a threshold to dare. A ghost train's tunnel. Flesh that opens like a lotus in flower to some, and to others a reminder of suffering, an archway to a death-like experience, the mouth of Jaws, or mother. And what about the clitoris, could it be the stump of a place where Djanggawul's sisters once grew huge penises? This female sensual organ is such a mysterious thing. It contradicts culture. It is primitive, savage, fragile and powerful. It has no rationale, it is not civilized. It is no wonder that historically its power has been undervalued. It may be that many women now are beginning to take an interest in their own sexual power, but like all human beings, when power seems to come easily, many women are doing exactly what men are accused of doing, forgetting the 'feminine'.

A woman's body is one from which mankind must be squeezed into the harsh reality of the world. Once we have left this powerful life-giving force however, we are on our own. Some women choose to remain childless, virginal or simply career-motivated. Others enjoy earthy sexual love with men or other women, and don't want to create offspring. Historically, the female body has become so fused with the sensual, such a deeply bonded image has been impressed into our memories and psyche, that a woman may become defensive and deny she is sensual or repress her own inner image. Equally, another woman may see 'sensual woman' as a certain stereotype, and feel inadequate in the face of such an icon. Others may manipulate the association of their body with allure, glamour and sexuality. These qualities are also labelled 'feminine', like the sensual, and are likewise projected onto woman herself. She may, just like a man, exploit her body, or use it for power or manipulation. A man too may see all women as sensual seductive mothers or whores. Equally, he may feel uncomfortable with his own senses, fearing criticism or sneers at his need for gentleness or imagination, both feminine qualities. He may sublimate his feeling world through artistic pursuits, passion for adventure, or many love affairs.

Every woman in the world will perceive her responses and her senses differently, for whatever she shares in body with other women, it is her own perception or experience, her own values and feelings which make her awareness of the sensual unique to her. Every man in the world will perceive his responses and senses differently, for whatever he shares in body with other men, it is his own perception or experience, his own values and feelings which make his awareness of the sensual unique to him.

Fruit or Seed?

Before writing was discovered it is hard to know how men really viewed women and how women viewed men. We can make guesses from the archaeological evidence – figurines and symbols are always offering us new insights – but we don't really know. There are many neolithic sites where female figurines have been found, and many controversies about what these wonderful carved stones and images mean. The female body was certainly a sacred part of ancient life, revered for its fertility and associated with the rituals of spring, divining, farming, child-bearing and sex. A stone engraving from Isturitz in France from around 20,000 BC has an image of a phallus inside a vagina.[5] Vulva stone symbols abound in many of the sites in Europe. Sexual magic, sacred love was the simplest and easiest way to celebrate life. After all, every time you engaged in sexual pleasure there seemed to be a connection to the birth of children, more sex, more sacred genitalia.

In the ancient world women and men's sexual anatomy seemed to hypnotize and enchant. A Japanese pottery figurine from over six thousand years ago has a highly glamorous pubic triangle and stylish adornment about her loins.[6] The 'Venus' stone figurines found throughout southern Europe may have been goddesses, mothers or lovers, but the one thing they have in common is their awesome femaleness. They are pure symbols of woman-ness and her carnal body. Phallic symbols also occur throughout Taoist, Eastern and early Indian mythology. In France, European archaeologists found a classic phallus notched with the phases of the moon.[7] (The moon does not always symbolize the feminine in other cultures, but it nearly always does so in European ones).

For the ancient Greeks, the anatomy of a woman's body was fearful and mysterious, but subsequently had to be explainable. Aristotle wrote extensively about the anatomy of the human being, thereby revealing the Greeks' own cultural confusion about passion, women and sexuality. Misogyny was rife, and it seemed that ordinary women were mostly contained. Eros was as popular as Elvis. Pliny, Aristotle and Hippocrates believed that men were solely responsible for the procreative force. Man's semen was responsible for the generation of soul, the woman being merely the provider of the body. 'While the body is from the female, it is the soul that is from the male, for the soul is the substance of a particular body.'[8]

Fear alone drugged men into a false sense of security about their seeding of the human race. Men can sow their seeds anywhere, at any time, but it seems to be the unconscious and sometimes conscious responsibility of women to chose and attract their mates. Women have to invoke the sensual or men would not hunt for the sensual. So women's bodies are fruits, they carry the seeds of future generations in the male children they conceive. But if a woman's body is now no longer a biological mystery, as it was once for those Greek scholars, why is it that many men still fear the sexual lifegiving force of a woman's body? The deep, swamp-cavern where lives the Hydra has long been a collective image of the womb. The image of woman seems to have embodied both the 'pleasurable' and the 'terrifying' for thousands of years. For the species to generate, the sight of the female body must arouse the male. The power of sexual potency must be engaged by a visual erotic connection to a woman, and this is where the visual sense of a woman's body is at its most powerful.

Western Eyes

Eastern cultures have always been more respectful of nature and its ambivalent forces. Yin and yang are the chaos, the coincidences in every moment that permeates all life. On the other hand the Western, Indo-European warrior originators discovered civilization and culture not so much as an afterthought, a kind of reconciliation for their barbaric potency, but as a reflector and antidote for nature's terror of which they were a part. When they discovered their own man-made power, the new civilized world simply went to their heads. In art, religion, literature and wisdom mankind could project the darkness and the light of the natural world. But however lofty and abstract the ancient patriarchal societies became, they could not get away from the one earthy reality that plagued them, the one thing they loved, hated, feared and desired: a woman's body. In her was reflected their own darkness, their own chaos, their own duality. So they covered her body and gave it a new name, 'wife and child-bearer', and 'she is my property'. But under the bed-covers she still lurked as a pagan memory. The carnal world still lived beside Everyman night and day to haunt him. This was 'his woman'. It was around this time, from about 4000 BC to 2500 BC, that early Sumerian and Indo-European culture discovered human sexual need was also a way to venerate nature's gifts. Taking pleasure in the physical embrace became both a sacred and a profane art, and the sensual was its greatest advocate.

NOTES

1. Quoted in Campbell, p.251
2. Hillman, p.23
3. The image of the body of mankind was suggested to me by Darby Costello.
4. (Plato, *Phaedrus* 247c) cit. Wright, p.125
5. Gadon, p.10
6. Johnson, pp.130, 131
7. Marshack, p.293
8. Aristotle, 1:1144-46

CHAPTER TWO

First Sense

*'When the panting horses of the rising sun first breathe on us,
there is the red of evening, Vesper kindles his late light'*[1]

A Sense of Creation

The early civilizations of Sumeria and Akkadia around 3000 BC were probably the first to record history. They were literate enough to leave evidence – art, facts and concepts that survived in fragments. In this sun-baked old world, the perception of woman was rapidly changing. No longer simply the cultivator, the protectress and the mother, she too faced the dilemma of whether to race with progress. Nature was changing with the construction of civilization.

The collective psychological metamorphosis was challenging nature. Mankind mined and quarried the earth for precious jewels, for minerals, nature's gifts and nuggets of beauty found in the earth that they could mould into shape. They tried to build towers that earthquakes could not move, block dams, control and tame the elements. Similarly they quarried women for clues as to the secret of her mysterious body. Mankind projected itself out into the heavens. Gods and goddesses filled our imaginations and we saw the stars in the sky as symbols of those gods and of ourselves. We began to try to find out how the world came into being.

Eurynome

In very few of the early Western patriarchal civilization myths does a female take sole responsibility for the creation of the world or universe. However, some very early surviving fragments of myths that pre-date the beginnings of the Sumerian and Egyptian dynasties reveal a different story. One of these earliest female initiators was Tiamut, who according to some Sumerian versions was not just

Mother Earth and the consort of Marduk the great god, but her menstrual blood was the origin of the fluid of creation. In Europe, the Pelasgians arrived on the Greek peninsula some time around 3500 BC bringing with them their own languorous initiator, Eurynome.

The story of Eurynome is one the most evocative of 'sensual' creation myths. Eurynome is perceived as the embodiment of grace, sexuality and darkness. She may be one of the first symbols, icons or archetypes of the sensual as feminine, the first psychic shadow of woman that emerged in Western humanity's consciousness. Here is the power of the senses, particularly when sexuality is involved, at its most erotic.

Eurynome is 'a wide-wandering dance'. She finds no place to rest on the ocean of chaos, as she dances eternally alone. Out of the west wind as it whirls across the sea she creates a giant serpent, Ophion. Likened to some pulsating, gyrating phallus he is chosen to be her consort, and as he coils seven times around her swaying body, she invokes the creation of the earth. Eurynome turns into a dove so she can lay the primal egg in the ocean. Now Ophion is ordered to coil around the egg seven times to incubate it. As the egg hatches, out falls all the things that exist, the stars, the sun and the moon and every creature.

But the most profound part of this tale lies in Ophion's growing arrogance. By being given permission to fertilize her womb, he now demands that he is the sole creator. His arrogance masks a deeper fear, for he knows how powerful she truly is. What did he find in this sensual, languid, dancing, wicked Eurynome? What did this great sea serpent encounter in her dazzling eyes? Was it exactly the same thing that a man encounters in a woman when he tries to dominate her? His own dark beast reflected in her eyes. It may be that the first archetype of fear and envy was sown by Ophion into this sensual creatress. The first sense of fear felt by Ophion sets the code for the continuing fear in mankind. Chaos, as personified by woman and

Eurynome, must somehow be ordered. The power of women's sexuality was never in question, but how to deal with such a creative and destructive force was.

There may well have been many distortions of original mythologies, as there was of women. As Jung comments, 'The most we can do is to dream the myth onwards and give it a modern dress'.[2] Joseph Campbell also writes: 'Myth is the secret opening,' through which the 'energies of the cosmos pour.'[3] We do not know, but we can guess, that the early connection of women to serpents, beasts, the moon, and Mother Earth herself, were all inter-symbolic of that which was not rational. Anything regarded as a feminine image, even the serpent Ophion who manifested as Eurynome's own creation, needed controlling.

Nature and all her manifestations were now perceived as non-masculine. The first zodiac was created in Mesopotamia, but because the zodiac seemed to move against the backdrop of the constellations in the opposite direction to the sun, it was deemed a feminine, dark image. The celestial belt was likened to a woman's nature, as dark as the new moon. Sensual, provocative, sexual, mysterious were all words that became attributed to women. Woman's body was mysterious, so mystery was woman.

Mankind was uncomfortably splitting himself off from the senses of body and soul, and running only to his new sense-perception of created time. His own sense of the sensual was becoming something to be feared, rather than to be enjoyed. The senses were considered instinctive primitive needs. It was easy to forget that the sense of awareness was the most important 'sense' of all. Myth making became a projection of mankind's darkness, and the gods and goddesses, now out there in the heavens, were no longer the powerful primal instinctive forces that men found so uncomfortable experiencing in themselves. These roaming, unpredictable archetypes of the unconscious were now consciously given names, labels, virtues or faults. They were called gods and goddesses, they became masculine or feminine, and were believed to have been created by the minds of men. Mankind now could be excused for feeling desire, for hating, for praying, for envy, for fear, by the parade of gods and goddesses whom they could propitiate. A whole pantheon of deities was born out of the human need for answers and explanations of their own experience. This was real emotion, real pain and suffering. The inner world of instinct, unconscious and shifting, became dream-like, elusive, and often dangerous, as humanity began to fall in love, hate and anger, to find he or she had warts or was rejected by a devouring mother.

If the Great Goddess ever did exist, then it was time to dismember her. Humanity's urgent need to always divide and define, to take apart and dissect was in its infancy, and the feminine image was one of its the first scapegoats. Men wanted to be masters, to be god-like, because the gods or goddesses were above mortality, they did not die. Whether men created an image of immortality or whether the god and goddesses immortalized the image does not really matter. What these early peoples did not realize was that archetypes form our perceptions, our perceptions do not create them.[4]

Inanna

In the early Sumerian and Akkadian civilizations the body was good, enjoyable, and many women were sacred harlots, or priestesses of Inanna, the goddess of love, fertility and life. These early Sumerian peoples worshipped the moon, held monthly rites to honour the lunar phases and every year a sacred ritual was performed. In the early days of Mesopotamia, it seems that men and women were equal in status. This was an era when sensual pleasure was a necessary part of the yearly cycles and rituals to ensure the growth and fertility of both crops and civilization. The most potent way to express the love of the sensual was through taking pleasure in one's sexuality, and it was here that sacred sensuality merged with profane sexuality and the goddess, Inanna, was known as 'true woman'.

Her mythology reveals the nature of sexual pleasure as sacred. Yet she was also the goddess of storms and of wars. A Tree of Life and sacred serpents attended her, bird images abound with her images. She was described as earthly, animal, yet carnal and sensual. She took many lovers, and her title as 'virgin' meant only that she was autonomous and self-possessed. Her consort, Damuzi, was, according to some sources, also her son and brother. Inanna was worshipped as the planet Venus, and may have been an earlier aspect of Aphrodite. The divine sexual embrace between Inanna and Damuzi was the heart of the Mesopotamian religion. A priestess danced and became a personification of Inanna, who with the King would recreate the sacred marriage to ensure fertility. Thus the life-force of the goddess was honoured to vitalize the earth.

> 'When with amber my mouth
> I shall have coated,
> When with kohl my eyes
> I shall have painted.
> A sweet fate I shall decree for him,
> I shall caress his loins,
> The Shepherdship of all the lands, I shall
> decree his fate.'[5]

First Delight

The cult surrounding Inanna celebrated woman and the harvest of the earth. There was no doubt that feminine sexuality was becoming closely associated with the senses. The erotic literature and art of the time revealed a pure delight in the physical body and its pleasure. This was the era of an upsurge in adornment and celebration of the body: the first fragrances, the first wickedness, the delicious spice of nature; dress, undress, veils, eyes and painted faces; incense, herbs, scents and oils; music, sacrifice, poems and song. Food was sumptuous, the tower of Babylon a sensual pleasure-dome, and 'love was all around'.

In Sumeria there was little conflict about gender or roles, and therefore men and women were less disconnected from one another. The sacred harlots of the temple had their own hierarchy, they danced, gave wisdom, or were sexual initiates. Their sexuality was demonstrated through ritualistic dancing, probably founded on the earlier, sacrificial Dance of the Seven Veils. Similar to belly-dancing still popular in the Middle East, these dances were probably once female-only fertility dances. 'Snake-like and vigorous hip and pelvic movements, the manipulation of veils, a descent to the floor and the ritual wearing of a hip-belt or sash.'[6] The sacred priestesses exhibited sensual and sexual messages. Their craft of love-making ironically imparted a different, more profound wisdom through the very carnal earthliness of the body as sacred.

What was the wisdom these sexual celebrations gave to man? Darkness, and the secretness of woman's body was available to share, to be experienced and to enjoy. Pleasure was as potent as intellect, and pleasure leads to love. To partake in that which embodies mystery, means one can begin to partake in one own's mystery. The divine could be found in the profane, the most profound yet eventually rejected secret of humanity.

The joy and sensual indulgences of these great civilizations however could not be sustained by love alone. For humanity, confrontation was inevitable. For more than anything else, man seemed to fear the power and the weakness of his own body. But where did that anxiety arise from? It may have arisen from the growing awareness, a new 'sense', of the unreliability of the body. One fundamental characteristic of human nature was gradually being revealed with humanity's consciousness, that life is insecure, and the sense of it more so.

FIRST SENSE

This Insecure Body

The more conscious we became, the more we realized that life is not a safe place for the 'egoic' sense. Mortality is finite and 'people die because they cannot join the beginning to the end'.[7] The divine could be found in the sacred place of body, but the body was unreliable and distinctly shaky, so how could the sensual world be a secure prop on which to lean all humanity's belief? How could the divine be found there? The weakness of the body had to be contained, isolated from the divine rather than fused with it. The denial of pleasure of the flesh could be a way of transcending mortality and life's lot. Humanity wanted what the gods had, immortality. The gods and goddesses invulnerability was unquestionable, so mankind became arrogant in the hope of exemption from the Moirae, the Fates who spin out our lives for us and cut the thread at the moment chosen.

The sensual world seems primitive and alien to our illusion of mind over body, steel gleaming before defenceless flesh as the surgeon's knife and sewing kit save us from the Moirae. The only sensual connection most people have with the body now is usually through sex. We attach ourselves to the erotic, both the sexual and emotional transformation of the senses. Touch, taste, sight sound and smell are still with us, yet we rarely attend to them, and rarely engage consciously in this basic sensory awareness. After all, to be sensually aware is the most profound sense of awareness of oneself just being, listening, perceiving, touching, smelling, tasting, experiencing. We are undoubtedly as insecure, or probably more so, in our bodies as we were four thousand years ago. We are eager to stand on firm ground, but the shifting sands remind us how vulnerable we are.

Sensuality, in the sense of sexual pleasure, revealed the limitations of the body. Taking pleasure in woman's earthly delight was no longer a safe place for the earliest civilizations, still wrestling with the secrets of the female body itself. Women's menstrual blood could not produce immortality, let alone longevity; it was still mysterious, but it seemed to be less magical than first believed. This first sense of mystery was gradually being explored, explained and rationalized. By defusing the fear of the unknown, mankind's anxiety about life and death separated our perception of these two senses even further, into life as masculine, and death as feminine.

NOTES

1. Vergil, quoted in Wright, p.35
2 Jung, 5, para. 337
3. Campbell, p.3
4. Jung, 5, para 337
5. Kramer, p.63
6. *Belly Dancing*, p.23
7. (*Alcmaeon* Fr2) quoted in Wright, p.65

CHAPTER THREE

Bodily Diversions

*'"The curse has come upon me!" cried
The Lady of Shalott.'*

The Sensual Curse

The human body has been rated nature's most exquisite work of art. It is understandable that both men and women have idealized and attempted to imitate it in abstract form, in sculpture, in art, in words. But because it is so unique we usually only enjoy its mysterious beauty through the pleasure of our sense of sight or the pleasure of our totally sensual-sexual world. But have we conditioned ourselves in an attempt to define and perfect the body? Have we shaped and moulded an imagined beauty out of a rationalization of what beauty ought to be? Nature is also ugly, cruel and distorted. Doesn't it produce handicapped children, elephant men and deformed beasts? Our ingrained sense of the delights and the horrors of the body stem from prehistoric times, just before civilisation vied with nature, when the female body was a mysteriously fecund, but also frighteningly secret and fearful place, and its most illogical manifestation: menstrual blood.

The most disturbing thing about menstrual blood was, it appears, its content. This visceral flow, at times black as night, then red and hot like lava, is messy, and sticks like glue to pure white skin. Much later, in the 1st century AD, preoccupations with menstrual blood reached a zenith of distortion. Pliny, in his negative description of menstrual blood and its effects commented, . . . 'to taste it drives dogs mad and infects their bites with an incurable poison . . . even that very tiny creature the ant is said to be sensitive to it, and throws away grains of corn that taste of it and does not touch them again. Not only does this pernicious mischief occur in a woman every month but it comes in larger quantities every three months . . .'[1]

Menstrual blood carries decayed sperm, uterine threads and black clots, not easily recognizable as once part of one's own existence. Mankind's splitting off from nature meant cleansing oneself from the cycles of nature. Women's mysterious blood was too close to these darker rhythms and it was best cleaned up. But it was cleaned up with an an already blooded rag. The menstrual cycle is the most basic, primitive cycle of being. A woman has no choice but to endure, cherish or ignore the blood flow. She can take magic pills and stop the flow altogether, but the choice leaks out uneasily.

For centuries menstrual blood was the magic essence of the life-giving force. It was perceived as nature's erotic discharge, ironically transforming women into creatures who are seductive rather than pregnant no-go areas. Menstruation is the most basic female rhythm, originally associated with the moon, and her cycle. Many sources suggest that long ago women menstruated in time to the full moon. "Moon blood' was thought by early South American tribal peoples to be the base constituent of mankind. [2]

Deeply connected to the cycles of the earth and the solar system, women became easy metaphors for all that goes bump in the night. The nocturnal wolf, wild cat, prowling evil creatures and fantastic beasts of our tortured imagination are 'feminine' images, and woman, the feminine's most potent personification. The magical 'material substance of generation'[3] was however the curse and manifestation of woman's dark sensual side. To bleed monthly with the moon also brought with it the unknown side of the moon. Emotion, feelings running high, PMT today, ambivalence and wild women then. A part of nature's twist we often can't come to terms with ourselves in modern culture. Women have inherited an archaic embarrassment about their periods. Echoes of traditional voices resonate in a woman's psyche about the dangers to men of taking sexual pleasure while women are menstruating. Could it have been a handful of brilliant scholars, like Aristophanes and later Pliny, enmeshed in cultural fear and its fashionable hatred of women, who hyped up this collective anxiety? 'For he (Ariphrades) outrages his own tongue with shameful pleasures, in the brothels licking up the spat-out dew, fouling his moustache, stirring up the scabs.'[4]

And do not forget the magical powers that were believed to be endowed in menstrual blood. Just one touch from a woman while she was menstruating could cloud a metal mirror, dull the edge of steel or gleaming bronze, and sour wine.[5] Imagine what it could do to the egg hunter-seeker and his sensitive organ! Some women still hide their tampons or pads in fear of masculine scorn or distaste, or maybe grin and blame the monthly cycle as a superb reason not to have sex: the power of menstruation works both ways, and modern woman can at least be creative with her blood.

Sense of Self

Before history was ever recorded menstruation and childbirth must have been a puzzle. Women bled for days and they did not die. They produced infants, their bodies awash with blood, and then some time later, they produced the mysterious substance again. This secret body was strangely pleasure-giving and taking, and yet pain-filled, at the same time. Man's first chip on the shoulder, his first sense of inferiority, may have come from gazing on the power of the life-giving woman. What was he and where was he in all this? It was easy to see the woman as mother, but what was he? A child yes, a son, certainly born of woman, but for what? Blood was intrinsically powerful, a positive and essential part of living. You would die if you bled too much and it came with birth. But maleness seemed to contain no mystery, a man's sexual organs dangled before the world and he was vulnerable. So something began to stir in his head rather than in his loins. Early man realized he had his own power. He could stop women's flow of blood. He didn't know why, but copulation seemed to work. The first pleasure in his own senses ironically meant the first separation from the senses. By believing he could control the flow of women's blood, he was unknowingly proving himself superior. This timing coincided with the first surge in humanity's sense of self. We became dynamic rather than passive, interactive rather than self-contained. The changing perception of relationships which urged on a growing need to divide and separate, eventually turned even the magical power and life-giving elixir of blood itself against women. The more aware of man's separateness from his imagined parthenogenetic woman, the more his boundaries and defences needed to be staked and claimed. Menstrual blood became an easy scapegoat for cultural needs.

It was not simply a collective drive for self-preservation that brought about the gradual split from nature and a new dualistic world, there was a glimmer in the 'feminine' eye too. If a woman was menstruating she could escape lusting eyes, she could reject 'love', and also cause emotional pain through her rejection; if she gave birth she had a unique power: she could give love and she could also postpone any emotional suffering, for the nurturing of a child at the breast may still the heart of love's pain, it is sensual, experiential, pleasurable. Then, if she extracted plant remedies and cures beneath the full moon, her healing power brought her lover back to her side. She could do both. She was powerfully wise.

Menstrual blood was dug deep into humanity's psyche as darker, more evil, more inexplicable than any other type of blood. Man could only deal with it by using the fear of menstrual blood and subsequently woman herself, to sustain his growing sense of separation from the womb, and from nature itself. Today, some New Guinea tribes still practice ritual incision of the penis to produce blood, known as men's menstruation.[6] The same taboos imposed on women are rigorously applied to men during this practice. Whether as a rite of passage, or a compromise with women to show men were equally able to produce blood without fear, this reinstated the ego's need for dominance.

Menstruation still carries with it taboos throughout the world. It is one that seems to be in reality man's most potent reminder of both his love and fear of woman. For the human race has asked woman to be the personification of sensuality, and for man to play the devil's advocate.

Menstruating Chair-persons

A modern-day woman seems to want everything. She may have bought into the myth that materialism will bring her happiness, and she may feel obliged to provide tenderness and compassion on demand. She has the power of preservation, so she loves her children. She rarely approves of the destructive power of potency, but has discovered the positive drive to do what she wants, to work freely, to live alone, or to be at the same time mother, lover and wife. But even the career-minded, business woman of the year menstruates. She has little choice but to accept this ancient earthly cycle which keeps her in touch with nature. So she sits in splendour, chairing the boardroom meeting, her tailored pin-striped suit and hair refined and elegant. But on standing to greet her business associates there is still pain, the grating, pulling cramps of women's viscera. She is both civilized and pagan, this boardroom enchantress. She has chosen the culturally constructed world of men in which to participate, but she, just like many men, may be denying femininity at the expense of power. For whatever underlying motivation or psychological complexes our chair-person has in becoming who she is, she is still just nature's child. Menstruating as she speaks, the taboo is only lifted because no one knows her secret.

Mother's Body

Women are fundamentally birth-givers: this is the female body's power of creation. But women nowadays don't have to give birth if they choose not to: this is consciousness. To be male or female on this earth immediately separates the human being into two different sexes. Whether we believe in reincarnation or not, when we are conceived, born, or whatever moment we decide upon for the life-force's mysterious entry point, we seem to come armed with certain weapons and with certain defences. We may receive conditional expectations from our society, we may interact with our cultural myths and our social and family romances to think, act and behave in certain ways according to our gender, but our temperament, our unique personality is inherent at birth. Jung wrote, 'The individual disposition is already a factor in childhood; it is innate, and not acquired in the course of life.'[7]

The child is first connected to the earthliness of life through the senses of its mother while in the womb. Mother is absolute, father is speculative. The certainty of who is your parent is almost 100 per cent with regards mother, but almost 0 per cent regarding your father. Father is usually a mystery to us throughout life. However much we love or hate him he represents for every human being the inner image of the solar, masculine principle, or the yang energy, within each child. Mother usually represents the lunar, feminine, yin energy. But father conveys and mirrors a mysterious sense of the stranger. We do not come from his body, we do not know him intimately as we once did our mother. Father is an unconscious stranger, who mirrors our own sense of strangeness, our sense of wondering, who am I? He can be either the welcome or unwelcome visitor in our mother's life, but it is the perception of father which constellates the journey to discovering our own mystery.

Mother's Sensuality

There is much said and written about our separation from mother. This first split from her body is symbolic of our first split from nature, and our first split from nature is symbolic of the umbilical cut. They are interwoven, these two senses of separateness, and they induce much anxiety in our personal and collective lives.

Our mother's body was our first experience of both dark and light 'sensuality'. We were cradled in her womb, given her blood, and either slept serenely for nine months or were jostled, disturbed and un-eased, depending on her emotional and physical state. Given that we are born carrying a unique entry visa into this thing called 'life', our parents, environment and relationships are merely the triggers for our own inherent reactions and behaviour. Our parents do not make us into psychological wrecks, they may help us on the way, but it is our perception of our parents, and the qualities we impose on ourselves and our relational world which make us who we are. Our inner image of Mother may not align with our physical mother, our inner image of Father may not align with our physical father, which presents us unconsciously with added complications for there is always this unconscious nagging dilemma that he might just not be our physical father either.

Yet mother is certainly trouble. She represents for us our birth and our death. And that's where fear creeps in. Once the cord is cut, our umbilical survival network suddenly has no say. We may scream and find the breast, which is sweet sensuality, nature's reward and warmth. But when our demands aren't met, or we are rejected, or dropped, or forgotten for a moment when the phone rings, it is the survival instinct that takes over. We must survive, for what is our fate? This is when the senses begin to cast their own internal shadows, leaking into power and manipulation even from the soft gooing face in the pram or the playpen.

Unconscious signals, discomforts, or obsessions which our mother particularly mirrors, leave early and uneasy tide-marks round our necks. Yet the warmth of our mother's breast, the touch of her hand as she stroked our fevered brow, these also are the loving sensual memories we carry. As adults we may perceive our mothers as housewives, nurturers, romantics or intellectuals, or we may have unconsciously cleansed our sensory realm of her fragrance, her beating heart, her emotional possessiveness. We may then favour relationships which are gilded with the same golden flavour as our image or perception of one or both parents. The creation and fantasy we have of what 'mother' means to us is projected onto our literal physical mothers, whether we are male or female.

Spookily, mother may actually resemble or act according to this inner image. We may carry an ancient mother image, an archetypal mother who has been unlived or unexpressed through our own mother and all the mothers of every mother before her. It matters not what our inner 'mother of all creation' image is like, but we may well see it mirrored by our own mother, and then by our partners, lovers, and women in general.

Jung described the negative mother archetype as 'anything secret, hidden, dark; the abyss, the world of the dead, anything that devours, seduces and poisons, that is terrifying and inescapable like fate.'[8] It may not be quite so awesome as this, but it can be. We may see our physical mother as obsessed with washing her tights, or simply cooking meringues at six a.m. She may be indifferent to our friends, or then again she may appear ready to gobble up every male friend we bring home.

The dichotomy for a male, however, may be emphasized by the different sex of his mother. How can the warm milk-giving mother whose breast one has fondled as a whimpering babe be at the same time a seductress and lover? As a fictitious 'John' reaches for the nipple, does the thought of sexual arousal also bring with it guilty thoughts about his own breast-feeding indulgences, or total fascination with the erotic imagery? Eros lurks even in a mother's embrace. The sensual experience of mother may be the most profound and most complex of all bodily contact. Freud may have been troubled by Oedipus, but he was wise about Eros as the force which keeps us vivid and unified. It is 'civilization' that is there as a 'process in the service of Eros, whose purpose is to combine single human individuals, and after that families, then races, peoples and nations, in one great unity, the unity of mankind.'⁹ Exploring the problem of the sensual mother is an image which recurs throughout history, and takes us back to the Great Mother herself, and her beasts.

The Mother Goddess was a potent reminder for ancient and pagan peoples that the world of the senses was ambivalent. The body became sacred, scooped out from the wildness and harsh landscape, particularly in the Indo-European cultures. Like the wild beasts of nature, terrifying and awesome, beautiful and dignified, the body became symbolic of life and death, simply because it lived and died. The beasts of nature were images of feminine and masculine qualities, and some, like the serpent have stayed locked in a poisonous embrace with the sensual.

The Serpent

The serpent was one such beast whose various relationships and affinity with women shaped much of our current assumptions about sensuality. Not just as Eve's tempting sidekick, but as a much earlier turn-coat, when men latched on to the snake's apparent mystical regenerative power. A creature that could shed its skin and be reborn became the symbol of eternal life and reincarnation. Originally considered a feminine symbol, once traditional patriarchal rule became widespread, the snake became a phallic symbol of potency as well as of seduction and charm.

Androgyny sneaks and coils around snake stories too. It may be that these very ancient creatures still exude a dank smell of the equality of men and women, when the sacredness of sexuality was never in doubt. Before Teiresius was blinded for his wisdom by Hera, but given prophetic foresight by Zeus, he had wandered across Mount Cyllene and discovered two snakes copulating. For witnessing a sacred act, he was punished by being turned into a woman. Later on he did exactly the same thing again and was turned back into a man. Teiresius owed his wisdom to both the witnessing of an archaic sexual act, and also to being both man and woman. Similarly, the feminine or yin energy, once acknowledged, takes us closer to nature's truth. Somehow the snakes are older than even our mythologies, they have slipped into our world to remind us we are sensual puritans.

Slinky

There is a notion that sensuality is about being slinky, a word often used to describe snakes and women. Slink literally means 'to creep' but is also an old word for an aborted animal, a miscarriage. Slinky women are miscarried women. In the eyes of Judaeo-Christian religious leaders, slinky women are those who are not carried to full term by Christianity's own pregnant dogma. It seems women creep like snakes, dowsed in their own uterine imperfection which must be aborted. Women as slinks are gelatinous and fluid, twisting, thickened and bruised as an aborted foetus. They are undeveloped, inferior, they are miscarried animals, not of mankind. Over many centuries this way of thinking gradually shaped the collective view of woman. In the 1st century AD, Paul stressed the inferiority of women: 'a man ought not to have his head veiled, since he is the image and reflection of God; but woman is the reflection of man . . . Neither was man created for the sake of woman, but woman for the sake of man.'[10] Then, in the 2nd/3rd century Tertullian wrote, 'woman is a temple over a sewer', and Plotinus suggested sexual and sensual suppression was a way to find God.

Slinky snakes live underground, they hide in holes in the darkness of mother Earth, they are Mother Goddess envoys, both her welcoming arms and her devouring smile. Snakes and the feminine are early interactive partners in mankind's body symbolism. Ninhursag, an early Mesopotamian goddess, was also known as 'Mistress of Serpents', and Ananta was the serpent mother in Hindu mythology, upon whom Vishnu rested between avatars. Kundalini is the powerful yin or feminine energy that lies coiled like a serpent in the base of the spine, waiting for sexual arousal so that this energy may be realized, rising to our highest chakra so that we may acquire divine wisdom. These early misogynists may well have reacted eagerly to sensual messages from their inner serpent, but they denied that the sensual serpent lived within them – rather it lived outside in the form of woman.

Artists and writers too have historically been fascinated by serpentine sinuous curves, or the coils of the boa constrictor as tight and as deadly as a woman's embrace. A painting by the nineteenth-century painter van Lutz, entitled, 'Sensuality,' evokes an individual perception of the corrupt woman and her serpent envoy. The denigration of women reached a tidal wave of obsession by the end of the nineteenth century, a repeat pattern of the Greco-Roman misogynists and the shark-infested waters of the fifteenth-century witch-hunts.

Medusa

There lies a glance of the darkest night. A sensual, coiling, spitting head of snakes, disturbingly erotic, and ominously renowned as a symbol of woman's hidden places. Another terrifying reminder of the secret vulva from which all men come and go. Medusa is an ancient deity, she could turn to stone any snake-charmer, except Perseus. She is the serpent goddess who symbolizes the hidden sexual power of women, and she may be an emanation of the earlier goddesses of the Nile, Neith and Ananta. She takes us back to menstrual blood again. To bathe in her gaze is to bathe in the source of woman's regenerative rhythm. The female uterine threads are woven tightly into her hissing serpent hair. Turning men to stone became a symbol of man's fear of his impotence, both sexually and physically. To be impotent is to be unmanned, defenceless, that old fear of the insecurity of the body transformed into a problem with power.

So as men began to recognize their paternity and lose their belief that women were solely responsible for generating the species, femaleness was still perceived as powerful, but containable. Civilization took the body into its hands and with a tiny brush and some coloured vegetable dye, it began to paint, dress and adorn the body, take pleasure in the joy of the eye as it recognized something sacred. The body itself became beautiful, engaged with constructed ideals and dressed with civilization's twirls of colour and embellishments of nature's passion. It was man-made beauty that encouraged the fusion of the sensual with woman. The art of the body crafter could heal, nourish and protect humanity from the eternal reminder of their fate – by the beautifying and adornment of fate's most potent symbol, woman herself.

NOTES

1. cit. Young, p.171
2. Chagnon, p.38
3. According to Pliny. However, Pliny also saw a gravely negative side to menstrual blood in his great work entitled 'The Natural History.' 24-79 C.E.
4. Ar. derides Ariphrades for inventing cunnilingus. Ar. 'Eq', 1284-6
5. Pliny, Nat. 7, 64-6
6. Tannahill p.44
7. Jung, p.38
8. Jung, part 1, para 158
9. Freud, Vol 12, p.313, The Pelican Freud Library
10. Paul, Corinthians 1, Ch 11

PART 2

Beauty Sense

'... for beauty is nothing but the beginning of terror, which we are still just able to bear.' [1]

CHAPTER FOUR

Skin Deep

So what is beauty? Dictionary definitions of beauty describe it as 'the quality that gives pleasure to our eyes'. It is also a value that generates sexual attraction and has been labelled a feminine quality. But beauty goes further than skin-deep man-made beauty. Man-made beauty is often illusory. Light and dark, beauty and ugliness are manifestations of how we perceive the world and accordingly judge it. Work of art, or aesthetic collusion, is easily constructed around the split between what brings us pleasure and what brings us pain. Beauty must align us to something considered as goodness, and it must bring delight to our eyes.

The word beauty comes from an ancient Sanskrit word meaning respect and 'gift'. It is this natural gift that hides beneath the veneer of mankind's constructed one. Humanity's fascination with separation and dualism has cast beauty and the sensual in the same viper's pit along with the feminine. They have become inseparable, and like Siamese twins they lie locked together in our minds, intertwined and interdependent. Beauty is woman, is sensual, is sexual. The words and the eye cannot but fall into the trap we have set for ourselves. The sensual is a quality which accommodates our sense of beauty, because sight sense-perception is our most powerful response to the external world. But what gives pleasure to our eyes can also be painful to us emotionally. What we believe to be beautiful may in fact have a sting in its tail. Beauty is both dangerous and vital. We have cultured pearls of wisdom by day, while cultured beauty stalks our night.

Beauty is a notion which artists, writers, philosophers and poets have for thousands of years attempted to define or measure. If we attempt to understand the meaning of natural beauty, we may be able to gauge why beauty and sensuality have become apparent allies exclusive only to women.

Natural Beauty?

'Beauty is only skin deep' and many other worn-out lines about this enigmatic quality lie like sour grapefruit in our breakfast bowls. We have flawlessness paraded before our eyes every morning, the beautiful child, the beautiful woman, the beautiful man. We are all meant to be beautiful. Beauty has become a neurotic commodity and another trading-post for collective wet dreams.

All fashion dictates and expresses the collective anxiety, growth or decay of the era. Currently, we apparently respond only to thin, gaunt, pale, child-like gazes. Beautiful people don't have big bottoms and large pendulous breasts, but they probably did in other ages. Take the 'Venus' figurines of the neolithic age. The norm was huge breasts, fertile and suffocating, flesh falling around a wide pubis, inviting, succulent and obese. These many images of woman some would find ugly, gross, distorted and definitely not beautiful. Our apparent contemporary love of 'natural beauty' is another 'con' in this constructed world. Many men and women consume the idea of natural products, plant oils, seaweed and mud packs to detox and enhance their city skins and office complexions. These are natural cosmetics with seductive messages encouraging us to become more beautiful, more sensual. But beauty is in the eye of the beholder, not in a jar; it is the most sensate experience we can have. This is a private sensual experience, for it brings us close to an awareness of that which emanates from all things, its innate nature, its mana, its own 'beauty'. A woman may 'shine' herself, a flower may 'gild' itself, a stone may 'lustre' itself, a seashell may 'burnish' itself. All are bestowing their presence, or the very essence of the object's beauty. The sensual experience of the observer 'sees' beauty, in other words, is touched by the quality of the object's givingness.

It is our individual perception that determines whether something is beautiful or not, along with social conditioning about what is beautiful and what is not. Conditional goodness and badness is infused into the object or face in the mirror because every age and culture expresses the notion of the beauty it desires. But true natural beauty is often ugly. The gifts of nature can be harsh, torrid and destructive, and this does not always give pleasure to our eyes. We have trained ourselves to respond to certain judgements about who and what is beautiful, for man-made beauty is ruthlessly discriminate. Only the part of nature which mankind finds unthreatening do we consider on the edge of beauty, the rest is banished to Hades' realm.

Yet natural beauty is the darkness *and* the light, the stones beneath your feet, the whirlpools and the slime, the scorching sun and remorseless hurricane. Natural beauty is indefinable because it is in everything. Beauty is both in the eye of the beholder and the beholden. It is the sense exchange of that which is offered by the nature of the object, and that which is accepted by the subject. Nature's beauties are both those of a rainbow, a sunset and a mud slide. Nature's gift is humanity, the individual spark of each of us, in all our extraordinary diversity of form. Nature's beauty is neither ideal nor fixed, it moves, metamorphoses. What we do when we sense inner beauty, that quality of the object which gives us pleasure and pain, is to take an in-breath. We become aesthetically aware. The word aesthetic is rooted in perception, which literally means 'to take', to catch, and 'to seize'. We seize the sensation, we gasp and hold it dear to our hearts.

Our perception of women's beauty has been defined by a blend of personal taste and cultural value ever since woman discovered she had another power over man by the way she looked, dressed and gazed upon him. It was inevitable that woman, dancing across the mosaic floors of Pompeii, would prefer golden rings around her fingers to nature's suffocating volcanic ash around her body. Who can blame her for siding with her painted face in the bronze mirror?

On the Eye

Some say that man's eyes and brain are directly linked to his phallus, which in primitive times may have been necessary for his survival.

Mankind is indeed an excellent voyeur. Prehistoric man learnt quickly to use his eyes for hunting, to develop his perceptive senses and look upon 'beauty'. Humanity's best contribution to nature is art, the design contact between eye and brain. This is also the prime sense which enabled men to take a dominant role. But what about our other senses? Historically, we have civilized ourselves out of the basic instincts and senses of our nature. Taste, smell and touch return us to the bestial nature of ourselves, separating us from the safety of 'rational sense'. Sight is the least tactile of the senses, the more able to turn into lofty thought, abstract, reasonable and unthreatening.

The eye of the beholder was able to craft, sculpt and shape beauty into whatever form it chose. For the ancient civilizations, building, founding, essentializing beauty meant escape from the fierce glass splinters of nature's beautiful broken glass. Beauty became man-made, a defence against the unknown, and shaping women to fit the image meant that women too were contained. It may have been during the early Egyptian dynastic period that the first fashion model was sighted. The early stereotypes on the catwalk were slim, wigged, long-necked women. Thin-waisted, small-breasted, and scantily clad for the dry climate, a woman was already seen as ambivalent, and subsequently mistrusted as these lines from an ancient love poem suggest:

'Of surpassing radiance and luminous skin,
With lovely, clear-gazing eyes,
Her lips speak sweetly,
With not a word too much.'[2]

The Minoan people may have been as difficult to live with as we are today, their psychology probably no different, but they glorified nature, women and loveliness. The Minoan goddess in her many aspects, coiled round by serpents, her head wreathed with poppies, or accompanied by birds and wild creatures, exudes nature's beauty and nature's compulsion. This was a civilization that revered the power of women as equal to that of men. Women began to value the art of beautifying themselves according to men's delirium. By evoking the kind of idealized beauty that men had actually constructed, women maintained an equality and harmony that became rare in civilizations ever after.

Women have to be perceived this way, as beautifully dangerous, sensually disturbing, otherwise men might never have attempted to civilize themselves. Men's own base, carnal and instinctive nature was, and still is, what they fear most. But they still see it only reflected in a woman's face. Constellated in a woman is the sensual, sexual darkness of his own nature. She appears to civilize him with her savage smile.

The penis is man's most primitive connection, arising spontaneously when the erotic trigger gets pulled. Sporrans and codpieces are not pretend power, they both shield and demonstrate a man's potency. Permanently opportunist, the male of the species will fornicate with whatever or whoever is at hand. This is the gender's instinctual and primeval nature, but because human beings have the power of consciousness men had to contain these primitive urges, and create morals, values and convention. Nowadays women are still covered up in societies where there is patriarchal fear and respect of the feminine, their faces and bodies hidden in Muslim countries, veils at marriage, religion still scorning the pagan reality of nature's beauty. A woman's beauty may be covered up behind man's unnatural dressings, but the wound still oozes uncomfortably through the sticking plaster. Contemporary men themselves may deny they are beautiful, or go to the other extreme and crazily construct man-made beauty around their own warts and pimples. Beauty is a quality of the sensual, but as embodied in woman it still weeps like festering scar tissue that has never properly healed, a reminder of the masculine carnal dilemma.

Male Dilemma

Men are highly sexual and so are women. Men can also be deeply sensual if they are in touch with their feeling world, but thousands of years of regarding the sensual as a feminine quality has meant that the average man still carries uncomfortable and unconscious memories of disparity between the sexes. Women too are taught that they are sensual and men are sexual. But again, the label has stuck for too long, and 'femaleness' has become a double-negative image of all that people hate most about themselves, whether it's weaknesses, anxieties, or just the dark.

Sensuality demands a response from us. It involves receiving, the involvement of sensation, of perception, of an in-breath from love, or a gasp from touch or warmth. The sensual involves sensate feeling, whereas sex demands only action, and a biological urgency. The biological difference between men and women creates a perception distinction in their relationship to the sensual world, and how they respond to it. This becomes the focus of much of our partnership *Angst*. Men biologically are conditioned to aim, direct and fire. In a primitive state, they can be aroused and quickly entertain themselves, then move on to a different partner. With enough space between intercourse primeval man could move fearlessly from female to female, or even man to man, without engaging his emotional feelings. A woman needs to involve all her senses because she is the one who must chose the right mate to impregnate her eggs, she must defend and nurture her ancestral fruit. Women, biologically, have to be sensual, aware, erotic and sexually powerful; they have to seduce, choose, tantalize, enflame and engulf. It may be that the only way man became civilized was because of woman's foil. Maybe she has to *seem* to be as *uncivilized* as he tries to *seem* to be *civilized*.

Civilization meant mankind could control its instinctive desire. Clothing our bodies and dowsing our primitive sexual instinct in intellect meant the urge could be repressed or denied. Massive or flaccid, men had, and still do have, a terrible problem with their own sexuality, simply because it defied everything they were seeking: progress. In any legend, myth or story, the hero has always to meet the monster or the dragon, the symbol of his own darkness, before he can find the pot of gold. The hero of mankind seems to have only just arrived at the dragon's cave, and he now seems to be facing the dragon. Discovering the secret of our sexual 'pot of gold' has taken thousands of years to even begin. Plato was the first to truly express his fear of his own penis as 'rebellious and masterful, like an animal disobedient to reason, and maddened with the sting of lust.' Our hero's quest may still only be in its infancy, but at least we are letting the dragon's blood drip on us with its magical message, that potency and receptivity are of equal value.

Beauty's Truth

Beauty comes to an end, it fades and dies like roses, like poets, like life. Beauty is an ambiguous illusion like life, its 'sweetness dependent on the loss of it'.[3] Beautiful women symbolize humanity's fate. For a beautiful woman may be able to give pleasure to our eyes, but she also menstruates and dies. This sensual beauty reflects our collective fate on this earth. Fragile, ephemeral moths around a hypnotic flame, we humans cling desperately to life. Clinging desperately to his potent life-force, man also finds the sexual power of a woman an uncomfortable reminder of his own sad lot every time he enters her.

The Greek philosophers, men of abstract beauty and thought, originators of reason, may have been the first unintentional denigrators of women, for all their respect for women and nature. St Augustine and other Church leaders, in their exploitation of Christianity, were to prove themselves originators of a more lethal misogyny. Whether through abstraction or transcendence of the carnal body it was the hope of these early patriarchs that they might rise above the awful quandary of their own physical lusting. Generating the species was essential, but the more seed he lost the nearer man came to death. The ancient fear still held.

In ancient China, the Taoist fascination with sexual potency as a way of extending your time on earth was enveloped with sensuality and love. The more you retained your semen, the longer you would live. Eternal life was as seductive and as tempting as woman. But orgasm was a risky business, it was a serious seeping away of your life-force. Did you allow women to take your precious life-essence, or adapt sexual experience as a means to transcend this earthly problem? The skills and arts of Tantra and other alchemical sexual practices in the East contort and twist sensuality into new meaning, but they embodied the skill and arts of sexual pleasure and more profoundly, the sense of numinous and emotional love.

In classical Greece, beauty became idealized. Both men and women were put into proportion. The beautiful boy became intellectually desired and the beautiful woman avoided. Women were deadly but exquisite, their beauty as thorn-ridden as the briar-rose. For the thinking elite, 'love as passion aroused by beauty' was being cleansed of women's dark reminders, and the love between man and boy, between tutor and scholar, idealized, conceptual, above the physical body. Physical beauty became Plato's complimentary ticket to understanding a different form of beauty, that of the perfect soul. Plato's ladder of beauty was rising a rung at a time above the ground of nature, but it triggered an enrichment and respect for the sensate world, as one which was a manifestation of the hidden gods or archetypes. This absolute and idealized beauty is not concerned with make-up, fashion or other cultural constructs, but with the essence of an inner beauty that radiates from within, a divine presence, and the means to awakening or 'gilding' the soul.

Resistance

Women's sexual magic is perceived as earthly powerful, an expression of the creative and destructive energy in nature. These emerging civilizations placed value on reason and purpose, on principles, ideals and intellect. Nature was no longer a place to find the divine. In the West at least, woman and sacred sexual practice was not the answer to transcend the darkness of nature's fury. Unconsciously or not, men had to contain nature's art. A force respected, yet feared, still lay beside him in his bed as his chattel, his wife. A woman at his side seemed to many to be as dangerous as the Harpies. The Harpies were notorious winged child snatchers, and vulture-like denizens of the battlefields. Originally beautiful winged goddesses, they became known as monstrous half-bird, half-women creatures. Their droppings were poisonous, and they stank of the decaying corpses they snatched from the battlefields.

Women were exiled to another status, along with nature and the Harpies. Haunted man could not escape their gold-tipped talons. Female sexual beauty was the one place man was pulled close to nature. The dark energy he imagined to be solely woman's and yet found so compelling tested him more than his war-ships ever did. Sex was necessary, but 'ever since man emerged from the dominance of nature, masculinity has been the most problematic of psychic states.'[4] In resisting sexual beauty, men were testing their own. To civilize the world meant civilizing himself out of his own anxiety. Putting woman into a conditional context so that he 'came to his senses', was mankind's mission impossible.

Classical Greece saw the stirrings of a different way of thinking about nature, about art and about women, but it struggled back to a sexual normality only by discovering that the power of sexuality as embodied in woman was in need of respect rather than denigration. Pederasty may have been rife, women may have enjoyed equality and suffered inequality depending on their status, but the emerging Hellenic Greece was a sexier era. Nature has learnt to be creative with mankind's twisted evolution.

Harmony

Beauty in Indian and Chinese ancient art carries a different echo, and follows a different thread. The weft and warp are compliant, flexible, centred. Still a cultural arrangement, Chinese Taoist art, poetry and belief is mesmerically sexual. Yet the symbolic use of male and female copulation in every art-form and every secret coded script, is both allegorical and erotic to most eyes, and highly suggestive of the sensual world.

Taoism is concerned with the harmony of yin and yang, the power of alchemical transformation, of living in the moment and participating in the Whole. Men and women, as starring symbols of yin and yang, have a part to play in the process of immortality and are essential for spiritual enlightenment. The world of nature is sacred, sexuality is sacred and although constructed by a male-dominated civilization, it seems that women represented the yin qualities that men believed they were lacking. Beauty here too became idealized at different points in Taoist history, but basically ancient Chinese philosophy is concerned with the sensual world as being the natural world. Fruit, birds, creatures and women all symbolize aspects of the wholeness and essential harmony of the universe. Throughout Chinese art sensuality is displayed as beauty. Here they interlock in a world of erotic magic. A few lines from an early Taoist poem express the idealized beauty of the time, and also the unrequited love between humans and goddess as the poet describes his encounter with the goddess of the Lo River:

> 'Bright eyes skilled at glances,
> A dimple to round off the base of the cheek –
> Her rare form wonderfully enchanting,
> Her manner quiet, her pose demure,
> Gentle-hearted, broad of mind,
> She entrances with every word she speaks.'[5]

In ancient Indian philosophy, the beauty of woman may be two-faced, as in Kali's dark, devouring nature and Parvati's gentle, compassionate one, but they are two aspects of one goddess and they are inseparable. These mythological deities were multi-faced, and had many emanations, beautiful goodness, as well as ugly, destructive and dark. Sensuality fused more easily into Indian civilization which, like Taoism, unscrambled the connection between the union of male and female into sacred sexual practice. Tantra and other mystical sexual initiations were born out of the sensual world.

Beauty may, as in any other society, have been idealized, a cultural reconnaissance of the era, but the natural world was welcome, and sensuality indigenous to the culture. Beauty was felt as an inner quality, and the divine in nature was sacred, unlike the idealized abstract concept of Western philosophy.

In Bhakti religious movements, women's feelings were given a high place in the spiritual devotions to Shiva or Krishna, and to humanity at large. The sensual world was honoured as beautiful, and both dark and light was necessary for erotic love. In the twelfth century a young devotee of Shiva wrote of her love for him, her awareness of beauty touched with sensitivity. This is the fusion of art, inner beauty and compassion. She *feels* beauty, rather than sees it.

> '. . . because they all have thorns
> in their chests,
> I cannot take
> any man in my arms but my lord
> white as jasmine . . .
> I love the Beautiful One
> with no bond nor fear
> no clan no land
> no landmarks
> for his beauty.[6]

This profound poem suggests that a woman could not take just 'any man' into her arms. Women are not nature's opportunists, they are her harbourers. Opportune literally means to lead into the port.

Biologically it is men who must enter a woman's gentle waters, but they must brave the treacherous currents to get to her first. The poem also suggests that men 'have thorns in their chests' and to embrace them can be painful if a woman does not defend her harbour. The lethal currents outside the calm of the harbour are like a scorpion's sting. This is its defence, and similarly a woman must sting to win her man into her safe waters. The young devotee who wrote this poem loves only the Beautiful One, Shiva himself, he who is not flawed like man.

Plutarch, the great Roman author, when writing of Cleopatra attempted to encapsulate the true value of sensual beauty: 'Her actual beauty was not in itself so remarkable . . . but the contact of her presence was irresistible . . . the character that attended all she said or did was something bewitching.'[7] Again, it is the 'feeling' that is evocative, the feeling and the presence of something mysteriously enigmatic. Art may produce exquisite form, abstract messages are sent to our senses, delightful reproductions of nature's wild or tender energies, but it relies on aesthetics. Men and women do not construct or produce sensual beauty, they embody it. A lover's sensual beauty can either be desire-making, or emotionally wrecking to us, or both. We are compelled by it, fear it and despise it, but we cannot do without it. Sensual beauty both engulfs and delights us with chaos. But it is our own inner beauty which we find constellated through the other. For example, a beautiful man may find it difficult to have a relationship with a woman. For every woman who meets him may be awakened to her own ideal of beauty, and she may have to give up part of herself to live with his beauty, which ingrained in her soul is a feminine quality. She may imagine him to be herself, and yet once her desire for his beauty falls away, she may struggle to believe she has any beauty of her own, and her wanting falls through the gap.

The classical Greek world founded an idealized female beauty, and this obligation became women's redemption. To be fashioned by cultural beauty meant you were more likely to take a husband, and marriage was highly desirable. Beauty became man's illusory muse, as did the rationalization of woman. Take Plato's attempt at de-sensualizing the female body, with the ghastly horrors of the 'wanderlust womb'. The womb, 'when remaining unfruitful long beyond its proper time, gets discontented and angry, and wanders in every direction through the body.' Would any woman trust this man to be her gynaecologist any more than he trusted a woman's womb? There, we are dancing with a great philosopher.

Age Before Beauty

This may be a worn-out cliché, but it seems to ring a bell with as deep and resonant a note as Quasimodo's. Quasimodo was not beautiful according to the cultural constructs of most civilizations. He was misshapen, deformed and an outcast. Western societies particularly have moulded human form into shape, reducing value in anything other than prescribed bodily statistics. Yet Quasimodo evokes feeling, he evokes a quality to which our feeling senses respond. When we are being sensually beautiful we are demonstrating our feeling world through the senses. We show how we feel about love, we touch and caress, we show how we are angered, we tighten our grip, or kiss passionately until our desire melts into liquid gold or divorce. The art of sensual beauty is to 'feel' feeling.

But what about age? Don't we have a written agreement, a code inscribed in dragon's blood, about an age when we are no longer beautiful? Quasimodo's ugliness is our fear of our own body's decay. Esmeralda's water becomes our redemption like our powder and paints. It seems woman pays a higher price for ageing, simply because in her beauty she embodies man's fate, and in her fading, changing, ageing skin she physically decays before a man's eyes – a reflection of his own decay. It is fear that produces cosmetics, these are our 'cosmos' or 'arrangements' which suit our collective eyes in our ordered world.

When women look 'haggard' and tired, wrinkles widening, eyes sinking, cheeks dry and bodies sagging, they are no longer considered beautiful. Women then are past their sell-by dates, unattractive, unwanted and alone. So if the grace of age is meant to come before beauty, why do we prefer beauty to age? Back to neolithic man and the problem with mortality. The root of all insecurity lies in the unreliability of the body and manifests itself mostly in the shaky realms of flesh. When beauty fades, the body is dying too. Attempts to resurrect beauty are poignant reminders of our greatest fear. Because we cannot contain age, we have given it a let-out clause, but we suffer our mortality and life's passion by cleansing ourselves with modern ideals of purity: pore scrapers, facials and fructifiers. Age devours man-made beauty fast – face-lifts fall, the hair thins and the skin cracks, as if in a sarcophagus. Devourers of mortal flesh, sarcophagi were stone tombs purposely constructed by the ancient Egyptians and Greeks to contain the corpse. The body's 'mortal coil' was respected and honoured, even in death.

Eternal Youth

Humanity has always had a go at brewing immortality, but never quite found the right elixir. The mythological hero Gilgamesh plunged to the depths of the ocean and found the magical plant which would give him everlasting life. Like any mortal, concerned with bathing and titivating himself before partaking of the magical transformation, he forgot the local sea serpent who could not resist the fragrance of eternal life and ate it for himself. The serpent was wise, Gilgamesh was in denial. If Gilgamesh could powder his nose and bathe in aloes and cinnamon oil, then the sheer delight of self-pleasure would see him through, until he must meet nature's beauty face to face in the eyes of the Hag.

It is said that Hags relish the death of those upon whom they gaze, so wisely mankind revised the Hag and veiled her so none would know his fate. Treasuring our youth is fashionable, and the

Hag has been an outcast since archaic Greece discovered the horrors of Hecate and the menopause. Idealized beauty became widespread because it is an illusion, a conceptualized man-made device to keep us amazed and amused until we discover what beauty really means as an inner quality, the divine gift within all things.

Age knocks hardest on younger doors, for once we acknowledge that beauty is as ephemeral as our own mortality, we can at least lie back in the bath and pamper ourselves with that knowledge, instead of panicking. Age does bring a tenderness to beauty, it mellows and knows itself. Fragrance, mystique, dress and charisma may be the qualities that bring pleasure to the eye, but true inner beauty gives lustre to that which it embodies. Aphrodite shines through those who honour her, whether they are twenty or seventy.

Man-made beauty changes with the great ages of time. Neolithic figurines suggest woman as fat, fleshy and laden for child-birth – survival of the meatiest was the norm. Was this beauty or merely nature's necessity? Fatness is often admired as a symbol of power in many cultures. In some African tribal peoples, status, prosperity, the health of the chief and his woman are dependent on being as fat as possible. Art reveals the ample bosoms, wide hips and huge bellies fashionable during the Renaissance and in Rubens' perception of women. As far back as Dynastic Egypt large ladies and well-endowed breasts appear in carvings in the temple of Queen Hatshepsut, one period of Egyptian culture that seemed less concerned with elegant statements of thinness.

Age can be held at bay for a while, beauty can mask the truth from us in the mirror, and we can use new devices to seduce, tempt and lure as easily as if we were naked. Whatever age we want to be comes in a bottle. But nakedness shows our physical age, and this is where a man may stop and hesitate, for can there really be beauty in the sagging wrinkled flesh of an aging woman? Hormonal changes are not conducive to contemporary idealism. Women dream of being beautiful and become depressed about their apparent lack of it, neurotic about the size of their breasts, or about whether to have liposuction or reverse the ageing process with the surgeon's knife. Beauty these days is the one quality whose image is burned into women's minds like a branded sheep. Ironically, for a few men such as the surgeon and the fashion designer, there is power to be found in the conditioning that woman must be young to save herself. From what? Immortality is for the gods, and although eternal youth never dies, neither does its imagined ideal and narcissistic psychology.

The Value of Beauty

But 'beautifying' oneself is a personal sensual indulgence, and there is value in its pleasure. The enjoyment to be had from flashing glittering eyes, painting cheetah's claws, moisturized bodies, sweet-smelling breath or fashionable clothes is a lure for romance, relationships and, most important, the love of oneself. In many Eastern, African, South American and other cultures men are more likely to be highly adorned and beautified than women, just as in the animal kingdom. But Western culture has exaggerated the fear of mortality, enmeshed as it is with the symbolic nature of woman. In one sense, the more conscious you are of your mortality, the more likely you are to separate yourself from what you perceive as death, an inherited philosophy that is deeply woven into our Western psyche. In other cultures, where nature or the Wholeness of the universe is a prevalent belief, life and death have formed a peaceful conjunction.

Self and Other

If we do not adore our own bodies as they naturally are, then perhaps by gracing our inner beauty with an outer one, we are also improving the quality of the beauty of mankind. Nurturing the skin is nurturing the skin of humanity, painting the lips is glossing mankind's mouthpiece and enhancing collective communication, bringing it shades of wisdom, colour sticks of new ideas and visions, new notions, new awareness of the kiss as the most sensual message of all. Adorning our bodies with fragrance, veils, mud-baths and jewels means we are celebrating the riches of the earth. We are adoring our species rather than denigrating it. The sensual message, mysterious yet inviting, then comes closer to desire and to love. If we believe we look good, we think good; if we send out loveliness, we may receive goodness and profound inner beauty. What goes around comes around.

So how did adornment, a powerful additive to woman as the living symbol of sensuality, palliate the crisis of mortality for both men and women, but also set up another deathtrap, a bewitching snare of sensual enchantment?

Still gold-threaded was his pillow like the braids in my hair.
I lay coiled across his bed, my thighs warm against the silken sheet,
my amber moon slung high in the desert skies. Here I waited for
ecstasy, for my sun-god and his golden iridescent barge.

NOTES

1. Rilke, p.92
2. Watterson, p.9
3. Thornton, p.143
4. Paglia, Sexual Personae p.125
5. *Chinese Rhyme Prose*, p.60
6. *Speaking of Siva*, p.131
7. cit. Lefkowitz and Fant, pp.150-1

CHAPTER FIVE

'*I have perfumed my bed*'

'I have decked my bed with coverings of tapestry, with carved works, with fine linen of Egypt. I have perfumed my bed with myrrh, aloes, and cinnamon'.[1]

Man-made beauty relies on adornment most of all when it is dressed for sexual dinner. Those grim reminders of mortality, our flesh and bones, are happily veiled by glitter and gold. It is not clothing alone that displays or enhances our beauty: bedazzling jewels, perfumes and body paint, symbolic belts, oils and hair pieces are all ways in which we express our individual desire to attract or reject. The nun's habit is a most explicit way of rejecting men, yet somehow, underneath the self-possessed Christian demureness hides a woman, with breasts, menstruation and womb. How can she not be aware of this body that embodies everything she seeks in a God? It may be that she forsakes her sexual sense for the sense of divine power itself. Split off from nature and projected into dogma, belief and ecstatic union with one godhead, the sensual is found through a spiritual abstraction rather than an earthly one.

Our adornment symbolizes who we are. It speaks our sensual language, offering others the chance to enter our space. Throughout history nature's bounty has been raided for its precious gifts so that we might offer personal value dusted with the treasures of the earth. These then become collective symbols of what is valuable to us. Similarly, by anointing ourselves with nature's adornments we are displaying our own unique value as individuals, for synthetic beauty relies on the creations of our imagination to give those of nature any human value. After all, nature itself is impartially wise.

Sensual 'Feelers'

Ancient civilizations may have first used adornment for magical talismans and charms. Evidence suggests that amulets, rings and necklaces were originally concerned with ritual, sacrifice and worship, before there was any conscious demonstration of taste, class or desire.

Unearthed in the peat bogs of Scandinavia, bronze age fertility figures from around 2500 BC are adorned with golden neck rings of the great love goddess Freya. Similarly, the necklace in Hindu mythology was sacred, representing the union of the darkness with the light. Kali wore a string of skulls about her neck symbolizing her connection with the dead, whereas the goddess Durga is adorned with pearls from the milky ocean, the light of the stars. These necklaces both protected the goddesses from other forces, and identified them. When we wear jewellery, make-up and perfumes, we are similarly protecting ourselves from those whom we do not want to attract, and identifying ourselves to those we do wish to attract. Attracting those who may respond to our individual sensual world, so that we may hook into each other and begin to relate.

Adornment is our sensual 'feeler' into the world. It is our unconscious affirmation of who we are and what we expect from others. The passwords may be different from culture to culture, but the meaning is the same. Adornment is coded sexuality. The cosmetics, combs, beauty spots and jewellery are secret amulets expressing our innermost desire. We tread softly through the boudoir seeking that which will embellish us. Whatever it is we search for in our partners, lovers or in life, our adornment often states our purpose more truly than do our words.

The Problem with the Fig Leaf

The fig leaf did very little to enhance Adam and did nothing for Eve's subsequent symbolism of woman ever since the book of Genesis was written. But were fig leaves the first sensual adornment in the world?

In other words, was Eve really the first woman to adorn herself? Hardly, but according to the Book of Genesis she was the first woman to cover herself. Strikingly seductive as the image of a fig leaf may appear, the symbolic fruit of the tree of knowledge offered to her by the serpent was to be her downfall, not the rather precarious presence of the leaf.

However, fig leaves have always been associated with sexuality and were early symbols of female genitalia. Figs were often included in love charms and were sacred to Venus. The fig tree was considered holy by the pre-Columban Mayan civilization. Figs were symbolic of human breasts, and were eaten as symbolic nourishment for fertility and sexual virility. In Eastern esoteric sensual arts the fig was a motif for the vulva, and the leaves its outer folds.

It seems likely that the mind behind the book of Genesis was all too aware of the suggestibility behind Eve's scanty covering. To be naked is not suggestive; we are blatantly vulnerable. By covering and disguising our nakedness, by playing with and adorning our bodies, we can invite others to us by our suggestive masquerade. Adornment is both a statement, and a disguise whereby we are invited to reveal the truth. The weakness of adornment is that its very mysteriousness becomes a target for those who wish to devalue the unknown. Equally, adornment and beautification mean we can become spellbinding. Women particularly became easy targets for those who would seek to devalue the mystery of nature, cosmos and soul. For a woman is the living symbol of all feminine qualities, including mystery.

Erotic Adornment

In early Minoan and Cretan art a golden seal ring from Isopata depicts the grace and sensual beauty of a goddess or woman naked above the waist, her hair flowing wildly free. Her skirt opens like a flower in bud down to her feet and sacred irises grow around her dancing form.[2] Bird-masks, apparent symbols of sexuality, are worn by dancers and goddesses depicted on bowls dating from around 1900 BC, and on a fine circular pedestal table, said to be ritualistic in origin, which stands in the museum in Heraklion.

A late Minoan fresco is embellished with exotic creatures, erotic crocus stamens fly through the air, and a seated goddess of nature receive more stamens from a mythical creature. The goddess wears necklaces of ducks and dragonflies and her extraordinary hair, some say, is a serpent. Adornment is symbolic, it plays with our senses. We must find a clue among the many to which we may respond, then we can be sure the message is for us alone. This is how we 'put out our antennae' as we sniff out adornment, see beauty, touch and caress oiled skin or perfumed hair. Adornment plays with our instinct and intuition.

Our finery and embellishments are metaphors and expressions for what and who we are. Whether these early Minoan images are depictions of goddess rituals, or merely comic strips of powerful women, their value lies in their expression of human interaction at that time, and our perception of it. Nature and dance are the 'inspiratrices' for our erotic imagination, and beauty, as always, brings pleasure to the eye so that we may banish the darkness for a while. Whether we prefer a fig leaf over our genitalia or the brashness of 20 carat gold on every finger, the message is the same: this is who we are, and this is what we are looking for. Beautifying ourselves is also an unconscious missive to another to sexually unite, or at least to try. Getting our kit on, instead of off, becomes the art. But it is our individual perception of and relationship to that message that is the profound difference.

Ideal Adornment

Worldwide adornment varies with culture and generation. The earliest known examples of fashioning and decorating the body, particularly the female body, comes from Egypt. Idealized beauty was born here. In vogue were small pert breasts, narrow waists and luxuriant hair, usually in the form of wigs. Egyptian civilization was constructed around its mythology, not its mythology around the culture. The ancient dynastic families lived their myths. Pharaohs and ancestors were incarnations of the gods, and the succession of the Pharaohs depended on the marriage of the Pharaoh to his sister or half-sister. In Egyptian mythology male deities only married their sisters, and the Pharaoh must do the same to claim right of succession. Beauty here was harsh, the sun cruelly disfiguring to the skin. Ideal beauty became ordered and symmetrical, refinement and civilization a foil to the rasping landscape.

The earliest man-made jewellery – anklets, bangles, bracelets and girdles – found in Egypt dates from around 4500 BC. But the earliest form of adornment quarried nature's own gifts. The cowrie shell was worn strung on a thong around the neck as an amulet to ward off the evil eye. It was also a symbol of the divine vulva, a powerful symbol of rebirth. Shells, ivory and tortoiseshell were commonly used until man made the first bead. As the Egyptian craftsmen threaded calcite, copper and hippopotamus bone beads onto the first necklaces, did they do so with pleasure, at the thought of beauty as they perceived it, a woman dressed and adorned to their notion? And had woman, insightful, wise and visionary, seen the copper threads in the dust, the rock crystal vibrating beneath her feet, earth energy embracing each shell or bone that she wore, and understood its power? Egypt's princesses, their girdles made of cowrie shells filled with fragments of metal, tinkled suggestively as they walked, alluring, provocative jingles of life and death in every step. Lines from the Insinger Papyrus warn men of falling foul of their own desire: 'It is in women that both good and bad fortune are on earth.'[3] Women represented fate in those days as much as they always would.

The Aztec prostitute was renowned for the care and polish of her body: 'She first looks in a mirror then takes a bath . . . She uses a yellow cream called axin to give herself a pale glowing complexion .

. . colours her teeth with cochineal . . . scents herself with nice perfumes and goes about chewing 'tzictli', clacking her teeth like castanets.'[4] Sex at a noisy price but an assurance that the whole delight would be accompanied by heady embellishments. Primitive man was being gradually groomed into civilized sex by the very woman whom he believed was 'nature's curse'. Yin weaves the silver threads of lunar rhythms through yang's furrowed brow and warrior wounds. A woman clacked her teeth, or flashed her eyes, she powdered her nose and perfumed her bed, she expressed desire, and demonstrated her specific needs according to the cultural sense of beauty at the time. Put an African violet in an igloo and it will die, but women and men have imagination, they adapt to one another.

The Art

We are responsive, but we must also decorate our responses, hang our bodies with fairy-lights, sequins and glittering gems. If we honour our bodies with ornaments, we are also honouring and celebrating the body of mankind. By adorning ourselves we are acknowledging the interplay of objects, beauty and nature's gifts with our bodies. Women don't do this any better than men, but they have become the adorned ones, unlike most other animal species. Pheasants, peacocks, lions, tigers, butterflies and many other species all parade masculine beauty to attract the female of the species. The Amazons, warrior women of Scythia, feared yet respected by the Greeks, were buried with swords, spears, jewellery and mirrors. Beauty eternally whispers, even in ancient graves. Without sacrificing her femaleness the Amazonian woman, strong and war-like, still faced the mirror and adorned her neck and arms with jewellery. What did they see reflected in their shining metallic mirrors? For the Amazons perhaps, beauty was another weapon in their armoury. We still use man-made beauty to disguise our less favourable features, if we have them, or to enhance our preferred ones. We still use cosmetics or fashion to defend ourselves and to attack others. Outer beauty and adornment is a beautifully crafted defence mechanism if we so choose. It gives us permission to be that which we believe we ought to be, and also to pretend to be something other than who we really are.

Inner or Outer?

Outer beauty does not necessarily correlate to inner beauty however. Female icons of beauty, including the supermodels and film stars who parade before us each day, may be fashionable examples in their own right, but their sensual beauty is no different from any one else's inner 'beauty' sense. We don't have to be beautiful in the eyes of fashionable and cultural expectations to embody and honour the sensual energy within, but it helps. Those who deny their inner 'beauty' may be losing a sense of their deepest connection to nature, and their interface with the stuff from which we are made, star-dust, gold-dust, earth and ashes. This is our beauty. Nurturing this body, honouring it with totems, gift, mud baths and oils, means we stay in erotic connection with the body of the earth and with ourselves.

Sheer Indulgence

Investigation of the early Greco-Roman cults honouring Demeter and Persephone has revealed the quantity and quality of adornment essential to their rituals and celebrations. Luxurious artefacts demonstrate their fascination with the sensual around them. An early vase depicts naked languorous women with beasts, pets, pot-plants and jewellery. Maybe the potter artist himself idealized the feminine, but the message of his perception of his world is evidence of beauty all the same.[5]

Etruscan women's sense of the divine in nature was powerfully honoured and propitiated. The early Etruscans were sensualists of great notoriety. Their adornment was dramatic, vivid and magical. Many of the graves of Etruscan women from the 9th to 10th centuries BC were filled with their personal items. Amber was found in large quantities, and it is believed Etruscan women used amber not only for adornment but for magic rites and spells. Over 3000 metal mirrors were found, 300 inscribed with love poems or messages. In women's tombs, rather than men's, magical amulets, mirrors, jewellery and gifts were discovered. One mirror back depicts Aphrodite in sensual embrace with Adonis – the pleasure of an older woman taken with a younger man was as serious a business then as it is today. Theopompus, a Greek historian of the 4th century BC was alarmed by the lascivious, luxurious living and general hedonistic pleasure of the flamboyant Etruscans. It seems the women, of the higher classes and aristocracy at any rate, lived in reasonable equality with their men. Theopompus was hardly amused by their sexual or sensual antics. Men 'shave their bodies smooth' and women 'take great care of their bodies and exercise bare, exposing their bodies even before men and among themselves.' He also noted there were many 'barber shops'.[6] Theopompus may have needed more than a shave.

The Renaissance saw another flagrant flourishing of adornment and beauty, albeit mostly through male eyes. Nudity became fashionable in art, but not in the flesh. 'We allow women who have a beautiful head of hair, a beautiful face, a beautiful bosom to show off these parts of their bodies; why are we so unjust to those whose beauty resides not in those parts but in others?'[7] One wonders to which particular 'parts' Valla refers. Could it be the supposedly fearsome, serpent-like vulvas of women, perfumed, groomed and oiled, suggestive beneath the foils of silk, gauze and lace, hidden and unadorned? By the end of the 14th century in Florence, laws were passed to stop women owning too many gowns, to ensure they didn't spend too much on ornamentation, and to keep them from having too deep a decolletage. Catherine de Medici reintroduced the fashion of using cosmetics: although used in many other cultures and in other eras, this renewed interest in adornment was a backlash against the austerity of the Christian church. By the end of the 15th century the ideal beauty had the whitest of skins, a high forehead, reddish blonde hair and no eyebrows – think of Elizabeth I and other painted ladies, the courtiers and royal barges reminiscent of Egypt. The power of woman's beauty compelled a re-emergence of secrecy, magic and alchemical cooking pots, and queens and royal ladies led the way with their artful skills. Illusions were created with lead-based powders – flirting with danger was closer to the skin in those days – rouges for the lips, and later patches, spots, false moles for shoulders, breasts and cheeks. The 18th-century courtier and writer, Emilie de Chatelet, wrote that wearing jewellery was one of her greatest sources of happiness.[8] Value was found in beautifying oneself, and thus identifying your criterion for worth.

Between the fifteenth and eighteenth centuries aristocratic women in the courts used their senses to survive. The female form became exaggerated, frames, stomachers, farthingales and bodices twisted, distorted and ironed out women's bodies into idealized images. The craftsman could now design bones, willows and wood around a woman's body. Woman was now constructed and framed, not caged, not quite like a wild beast, but certainly containable. From the mid 14th century onwards man could ensure her chastity belt was firmly locked in place when she wandered. Man's own embellishment was the woman on his arm, and the key to her serpent power locked in his codpiece.

The delicate subject of woman's sexuality gradually became a collective obsession. By the end of the nineteenth century *fin de siècle* art and literature relied heavily on the idealization and demystification of woman's 'parts'. What was she, who was she? Daring seductress by night, virtuous wife by day. The more adorned and dressed, the more daring or outrageous a woman was, the more the image of woman as sensually provocative became widespread.

BEAUTY SENSE

The Scented Path

Perfume is the most suggestive of adornment. To enhance the female pheronomes is an art that requires much trial and error. Females have a biological scent that attracts, to let the predatory male know of her presence. This may be less obvious in the human species than any other, but it is still a sense given out and a sense received. In the wild, animals have a better developed sense of smell than man. We have lost touch with our nasal sense, and nowadays odours and powerful body excretions, like anything else which appear to suggest the unpleasantness of life, are reviled as negative. The female dog gives off the most exquisite smell if you're a male dog. Every dog worth his bark will be magnetically drawn to this canine aroma until he drools. Women have to make more effort where fragrance is concerned, but a woman apparently does exude something around ovulation time that men unconsciously cannot resist. Similarly, men who use deodorants and after-shave are assuming that they are displaying their attraction, or is it their detraction?

Man in primitive search for a mate needs to cover his track, hide his own scent from other predatory males so that he can get to the 'hot woman' first. Like dogs who roll in fox's urine or bird-droppings to disguise their scent, man rolls his deodorant under his arms, his stick of perfume across his shaven face in the hope he won't smell of himself, just long enough to baffle the rival. The competition is confused, temporarily defeated in the attempt to pick up the scent of beautiful woman, and the embittered loser turns to misogyny or the brothel.

'Beautiful woman' herself loves to play with smell. She is in her element in bathing, decorating and scenting her body. Ancient Egyptian women were renowned for their sweet scent as they passed by. Scented unguents like olibanum and terebinth were poured over rich queens and princesses of the dynastic periods. Cones of perfumed oils were clasped to their wigs so that during celebrations and feasts the waxy oils would melt down across the hair with alluring scents, disguising the less favoured pungent smells of food, BO, and bad breath. When the cult of Isis became popular in Roman Italy, women sprinkled the road as they walked, with bottles of balsam and other perfumes, and everywhere, the world celebrated the experience of fragrance. To be scented was to be sensual.

Some men sniff women's undergarments like sexually tormented celibate monks. Circe knew how to handle Odysseus's leering, debauched crew as they made drunken assumptions about her 'island hideaway', so she turned them all into swine. The power of attraction is what adornment is all about, it is what our sexual messages, both pheronomal and sensual, relay to others. The knicker-sniffer may get some temporary joy, but he is missing the pleasure of the real thing.

The Purpose of Adornment

Whether women are like Eve, feeling obliged to wear the fig leaf to disguise and yet suggest their erotic and sensual nature, or like an ageing film star, merely covering up their face-lift scars with powders, the purpose is the same. To adorn oneself is to enhance one's potential for attraction, not to demean it. With magical pots of paint, with plant-extracts, dyes and chemicals, with precious gems, crystals, talismans and cloth we are all dressed for our sexual dinner. We lay ourselves with care as we lay our supper table. Those whom we invite to us must be the best possible choice for the sexual union, the food of creation, we are about to share. We also adorn ourselves for personal reasons, to make us feel better, more worthy, more valuable and to improve our self-esteem. However, the problem is not in the offering, the attraction and the response, it lies in the darker shadows of self-deception, misunderstanding and power play. We may adorn ourselves as well as any lyre-bird decorates his bower nest, we may gild and perfume our beds to enhance our mystery, but we are split off from our inner sense of beauty by the very consciousness that we must endure. It is our fate to know.

It is our fate to be in relationships and to bear it all. Woman, as the archetypal embodiment of the feminine, carried too many reminders of humanity's own despair with mortality for her to get away with the license to chose her mate. In learning how to attract through adornment, her beauty became both her salvation and her curse.

NOTES

1. Proverbs, 7: 16-18
2. Johnson, p.148
3. cit. Watterson, pp.13-14
4. cit. Tannahill, pp.306-7
5. Fantham, Foley, p.240
6. Fantham, Foley, p.248
7. Lorenzo Valla, cit, Tannahill, p.284
8. Anderson, Zinsser, Vol. 2, pp.17-20

CHAPTER SIX

Girdles of Desire

'What the Imagination seizes as
Beauty must be Truth'.[1]

The Trouble with Aphrodite

Adornment is our means of attracting and repelling, and also our demonstration of an identity. If a woman lifts her breasts with underwire, dabs patchouli scent on her neck, bathes in musky oils and hangs jewels around her arms, this is an expression of who she is. If men conform to cultural fashion, or display eccentricity in hairstyle, dress or fragrance, they too are displaying a deeper image of themselves which they may prefer not to express in words. What we say often does not correlate with what our deeper intentions are. Our bodies cannot hide the truth as easily as language, however much we cloak and dagger our sensual script. In intimate relationships, the language of this feral, erotic world has usually only one message: 'I desire you to desire me'.

Aphrodite is the most symbolic goddess of sexual attraction. She is both earth and air, paganly meaty and yet civilized abstraction. We shall dance many times with this one yet. Understanding her dual nature may help us to understand the dual nature of love, and how woman's own nature is perceived as a perpetuation of this duality. Aphrodite moves within us and without us, for her power is equally potent in nature and civilization's beauty.

In her earthly emanation Aphrodite is tidal wave and mountain stream. She is 'beelike', her honey sweet and alluring, but her sting deadly to the allergy-ridden business exec. Beauty and attraction, adornment and sensual pleasure are hers to enable her to seduce whomever she pleases, from the sexual teetotaller outside the massage parlour to the fitness freak at the gym. Wisely she will chose only those who like the taste of her honey, those who will respond to her language. This is where Aphrodite represents our need to come out of ourselves and be in relationships, to refine and compromize our instinctual world, the natural, monstrous, primal side of ourselves. She is as paradoxical as we are. All human beings must grow out of base instinct into cultivation so that we can survive with others. The trouble lies not in acknowledging the animal within us, but in suppressing or denying those basic instincts or making excuses for the overt exploitation of them. We may have problems relating with the world, not knowing the true value of our senses, or finding they conflict with society's or family's imposed values. We have to learn to synthesize nature and culture or we shall be in danger of 'sacricide'.

Aphrodite embodies the earthly delights of sexual pleasure. She is the sensual persuader supreme, attended by Peitho, the goddess of persuasion. Whether in an office lift or on a commuter train, she can erotically fascinate and taunt, tease and weave her spell, binding men to women, women to men, brutally sometimes, painfully, yet always inevitably. Yet here is the paradox, for Aphrodite also civilizes the instincts. She finds us another upon whom to reflect our inner beauty, so that we can discover how to live in relationship with others and ourselves, to be in the world and part of it.

The erotic trigger becomes Eros's domaine – once fused by desire, Aphrodite must ground the abstract notion into human passion or love. She personifies both the ideal of beauty we carry individually and also the earthy, visceral, dark decay of our erotic beginnings through her envoy Eros. Aphrodite enables sexual attraction to materialize into pleasure. She watches and attracts. She is both nature and culture embodied in sexual power. Her earliest worship is rooted in Ishtar and Astarte, and as the oldest triple-goddess in many disguises, she was eventually assimilated into the Greek pantheon as a more civilized deity.

Dazzling

Asked to judge 'who was the fairest' out of Aphrodite, Athene and Hera, Paris had little choice when Aphrodite flashed her erotic girdle before him. Knowing her beguiling girdle had smitten Paris with an urgent sexual passion, Aphrodite offered him the most beautiful woman in the world, Helen, as his wife. Helen is discussed later, when we lose beauty to sexuality.

Our own girdles of desire, like Aphrodite's, are purposeful and potent sexual arousers. We become irresistible when we wear the erotic envoys. Each of us has our own, individual pattern and code to attract one man or woman at a time, or more if we are so lucky. Our girdle of desire is the sensual breath that we breathe across each possible mate we meet. Aphrodite brazenly dazzles her girdle across her breast just as Eurynome weaves the phallic serpent coils around hers. We are using our sense-perception for exactly the same sexual purpose. This is where our inner serpent awakes, ready to strike or to lure. We are both the snake and the snake-charmer. The ambivalence of Aphrodite is in us all, but it is woman who has been her most conspicuous vehicle.

Snake Charmers

By the end of the nineteenth century sexuality was so dammed up by the civilizing process that wet-dreams seeped out into literature and art. The old images of the serpent became phallic as humanity became more civilized. Woman seemed to twist a graceless coil of evil around her man, as the serpent was seen as both temptress and phallus. The erotic charm of a woman was according to Flaubert corrupt and bestial, as sinuous and cunning as a python. Flaubert's character Salammbo, not unlike Salome, is a priestess of Baal. Flaubert describes her encounter with the python, her writhing dance suggestive of sexual desire. The snake-charmer and man-eater par excellence, she is a vision of female intimacy that fed Western artists and writers hungry for pythonic *digestifs*.

'. . . the python fell back, and putting the middle of its body round her neck, it let its head and tail dangle, like a broken necklace. Salammbo wound it round her waist, under her arms, between her knees; then taking it by the jaw she brought its little triangular mouth to the edge of her teeth, and half-closing her eyes, bent back under the moon's rays . . . it tightened round her its black coils striped with golden patches. Salammbo gasped beneath this weight . . . with the tip of its tail it gently flicked her thigh.'[2]

Erotic literature hides behind symbol and euphemism. Flaubert's confusion is understandable. The woman as snake-charmer or the woman as snake? The snake as phallus, closing in on her parted lips, or her own triangular yonic symbol close to his mouth? The sexual connotations are tremendous, as is the literary device for Flaubert's own sexual complexes. Yet he must have sensed, he must have smelled that both men and women are at once the snake, the erotic trigger and the moon's rays, we are gaspers, lovers, coilers and predators. To attract is to cause chaos, for only out of chaos can order come. When opposites collide there is an oscillating union.

Women and men collude in the dynamic tension between them, yet we are all scattered with the same ancient dust settled upon our ancestors' tombs. A woman offers the power of preservation of the human species as she inhales the natural world and then breathes it out again. Her body is rhythmical and in harmony with the universe. A man has to be attracted by the femininity and femaleness of a woman, because he has been distracted from his own 'feminine' qualities by his 'egoic' sense of domination. A woman cannot let a man get away with hubris if she is to procreate.[3]

Civilized Attraction

All sexual imagery, all erotic turn-ons are a fusion of intellectual attachment and body response. They are the result of civilizing 'attraction'. Locked in our genes are millenia of images to which we respond. We have cultured our senses to watch for certain messages in the bodies of others: a knowing wink, a come-on smile, a look of passion. The more we use these phrases or sexual jargon the more we seem to devalue the very purpose of the attraction and the game. The sensual caress, the touch, the charm, are about giving love, as well as receiving it. The more we stay in erotic contact with each other, the more we are in erotic contact with the world.

Eros and erotic energy have been distorted and devalued by the world of pornography. Like any other archetypal quality, the 'erotic' is corruptible. Eros is a greater dildo than the sexual vibrator. Dildo is a word of uncertain origins, but it may stem from the Greek word meaning to expand, to widen and enlarge. The dildo of erotic living is about sensual living. It is about opening up to change, to shifting times, to cycles and transformation. It is about being aware of what we are sensing, not just physically, but in how we react to others. Who is it who gives us a sense of direction and purpose, what gives us a sense of fear, where do we feel a sense of love? What is it about the way I look that gives others a sense of pleasure or repulsion?

The sensual includes our relationship to sexuality, but is not necessarily reliant on sex. Attraction and beauty are there to incite and to lure, and without them the erotic process of nature would not endure. The erotic is not just about sexual excitement, it is about life and death, our attachment to our bodies and pleasure as a means of coming to terms with our very mortality, there in the arms and the love of another human being. To debase eroticism is to deny humanness. Pornography uses Eros as its trigger just as any other expression of life, but unfortunately it claims ownership over the erotic which it exploits for sexual gratification. This is not love, nor erotic beauty and transformation, it is civilization's sickness. However, sickness is usually a symptom of a deeper 'dis-ease', its manifestation enabling one to see symbolically what truly is in need of a cure.

Women have usually been seen as the protagonists in the mating game with all its twists and turns, its labyrinthine love an alien place for Saturnian materialism. The only way it seems most of us can love one another is to be first ensnared by each other's sexual power. Intellectual love and soul-love may come later, or straight away if both partners are aware enough, but somewhere along the line in men there is a conditional fear of fate, castration, mother, and all that these images symbolize, and an inherent unconscious belief that women are responsible for that fear. This seems to have originated in the millenium devoted to one God, the Omni-Potent guy about town. Until now, that is.

Nowadays, Aphrodite is becoming our friend again, and like the Greek poetess, Sappho, we may fear and love her equally, but in that we may see the paradoxical necessity of our different sense-perceptions. Aphrodite asks of us to honour the erotic, to honour the earthly need for human touch and the abstract need for ideal love. Aphrodite, both the goddess and her qualities, are intrinsic within all of us whether woman or man. The feminine principle in men is as clouded by cultural expectations and illusions as the masculine principle is in women. Sappho, possibly lesbian, possibly bisexual, but certainly a worshipper of Aphrodite, had deeply personal feelings and visions, whether about men, women, archetypes or gods. Sappho's writing suggests that gender is irrelevant to unconditional love. In one of her many fragmented poems about Aphrodite she says:

> 'Let the depths of my soul be dumb
> for I cannot think up
> a clarion song about Adonis
> for Aphrodite who staggers me
> with shameful lust
> has reduced me to dull silence,
> and Persuasion (who maddens me)

from her gold vial
spills tangy nectar on my mind.'[4]

The shifting, changing nature that we perceive as feminine is Aphrodite's nature, deceptive yet pleasing, beautifully beguiling but dangerously chthonic. Attraction has to induce chaos, it has to agitate and entangle a man so that he responds. Aphrodite in every man has been projected historically onto woman, so a woman becomes more like Aphrodite with 'every breath she takes'. If a man's sense of his physical instinctive nature is so monstrous, he may have to be forced into feeling 'feeling', rather than just feeling desire. This is why he may deny feeling and take on abstract thought as his visor. The man in the iron mask might secretly have worn a velvet hood, but it was Aphrodite who sewed the last stitch around the eye holes.

Beauty in Transition

Aphrodite is the most ambiguous of goddesses because she tests mankind as we move and fluctuate between nature and culture. She tests Psyche's mortal beauty through her love for Eros, and similarly she was the major cause of the fall of Troy when she awoke Paris to the beauty of Helen. Helen was in one sense Aphrodite's emanation on earth, and the beauty of mortality was to be tested for over a thousand years. Was abstract elegance, the constructed symmetry of the Greeks, really of value? Would the flow of life, death, rebirth, nature and cycles of the cosmos survive mankind's new plaything, beauty?

There is a psychological notion called 'the transitional object'. In other words, what we attach ourselves to when separating from the umbilical cord and mother. It is something in life we must grab hold of, whether it's our book, Bible or cigarette packet. As a child it is usually a toy, a comfort rag or a dummy. Later on in life we may shift the object of our attention to a cigarette, a drink or another lover. Life itself is a transition between the arrival and departure lounge. In the cosmic scheme of things, humanity's transitional object is the collective mind's vision of constructing beliefs, values and concepts, constructing the illusion of reality, constructing beauty so one can escape from nature's irrationality. Mother really is our life before we find our personal transitional object, we cannot exist without her, but if she is our life then she can also be our death. The individual's personal transitional object may then become the fantasy world itself, for the world of dreams may seem safer than the unpredictable arms of mother. For example, a man may fear any kind of sensual attraction. The earthiness that some women display may be uncomfortable to him, hidden echoes of fate and death possibly triggered by his perception of mother as physically demanding, smothering or monstrous. The world of the abstract, of mechanics or materialism, may become his own transitional object and anything resembling nature's fecundity or exotic beauty a devouring trap. Thus a modern-day Helen of Troy would certainly be perceived as dangerous to him.

Helen was an essential kill-joy, a reminder that beauty, not manmade beauty, but the inner beauty of Aphrodite's nature, is double-edged: mysterious, deathless, lifeless; life-giver, life-taker. Helen of Troy was archetypally dangerously beautiful, and like Aphrodite she dazzled. As Aphrodite's device, she was a transitional object sent to eroticize the course of history. The 'face that launched a thousand ships, and burnt the topless towers of Ilium' was a symbol of supreme sensuality, a woman of wiles who embodied earthly pleasures.

Helen

Legendary beauty dissolves the reality of it. Finding visual examples of Helen in Western art is almost as impossible as imagining how one woman's beauty could have such a powerful effect upon Western humanity. Yet Helen had to remain an enigmatic mysterious beauty for her ever to be so alluring, so sensual. Once a visual image is created in a painting, portrait or sculpture, it remains in our minds – look at the Mona Lisa or Botticelli's Venus, or rather try to think the words without seeing the pictures. We freeze an icon in a timeframe, and it is caught *sub rosa, in camera*. Written about in poetry and prose from Euripedes to Marlowe, Helen's own attraction and beauty changed with woman's shifting identity and culture's consciousness. It was safer to write about her beauty than paint it. So who is she? And why was she so dangerous to portray in Western art?

Helen was in most accounts half immortal, although she may have been assimilated into Greek mythology from an earlier moon goddess cult based at Troy. In the Greek pantheon her father was Zeus, and her mother Leda. As a semi-immortal, she was above human morality, although not mortality, and like Aphrodite she was attraction personified. Helen's beauty was renowned, likened to that of 'golden Aphrodite', when she chose the handsome Menelaus as her husband out of a medley of Greek heroes. There are many conflicting versions of her character, first as manipulative, betraying, and seductive, then later restored as virtuous and true, mistaken and wronged by a ghost-part Helen who took her place at Troy. But whatever the case, the Greeks went to Troy because of Helen's beauty. Paris may have been a fine lover, Troy may have been a clever wheeze of Aphrodite's, a retaliation against those who did not propitiate her power, but the former suitors of Helen, dashing dynamic, potent heroes who had sworn an oath of allegiance to Helen, were summoned to Troy and war. Men died and killed for her beauty, in more ways than one. To the Greek philosophers and playwrights woman's sexual beauty seemed to be the true source of man's emotional and physical self-destruction.

In myth Helen had three, possibly four, husbands. First Menelaus, second her lover Paris, and then after Paris was killed she married his brother Deiphobus. At Troy she was betrayer or hostage, but there was no doubt that her enigmatic beauty instigated both war and wanting, the very things in men which seem to break wind most often. A dialectic flatulence between desire and confrontation with its resultant end-product, emotion.

According to later accounts by Greek playwrights, when women were obliged to enter marriage and patriarchal misogyny was at its most rife, Helen apparently went on to seduce Achilles who was 'whirled in his dreams by her perfect body'.[5] But whose idea of a perfect body was this? A cultural definition of perfection, as symmetrical as the pose of the 'Kouris' the sculptured embodiments of the perfect boy of the Greek imagined world. Apparently Helen and Achilles lived in eternal bliss, confirming to the Greek misogynists that the contract of marriage was the only solution for containing dangerous women and their wily tricks. To exaggerate the power of women's attraction demeans that very power. In the ancient world women had little chance to redeem themselves if they were all thought to be Helen of Troys.

We now find Helens everywhere, but the new Helens are women who are losing touch with the quality of sensual beauty, and using only the dark side of earthliness on the road to equality. We may be sensual and pleasure-giving, lovers and mothers, but to choose only the path of sexual sensationalism and eroticism, means denial of the other qualities of femininity in woman. The archetype of Helen is an easy hook on which to hang one's own power issues. There are usually many underlying motivations for our struggle with another's beauty, power or grace, but it is usually around our own power, beauty and grace that the problem really lies.

World-Wide Attraction

Naevius, a comic playwright of the 3rd century BC, has left some fragments of verse. Here he describes a dancer from Tarentum. 'She nods to one man, winks at another, caresses this man and embraces the other: her hand is busy over here, she stamps her foot over there, she gives her ring for one to admire, and entices another with the pout of her lips. While she sings with one man, she writes messages with her finger to another.'[6] Aphrodite channels seduction through the body of woman.

Less anxiety-ridden cultures than the West found a deeper meaning through the sexual sensuality of the body, and particularly the rhythmically attuned woman, ancient symbol of the feminine, the moon and creation. Early Indian art, so infused with mythology and religion, carved erotic figures in friezes around many temples. Statues, carvings and inscriptions suggest this was done not only for the love of sacred sexual persuasion, but also to adorn and beautify places of worship with nature's dancers, weavers of love both divine and mortal. The statues at Khujaroo are sensual (i.e. conveying a sense of seduction) graceful and elegant in their portrayal of the erotic fusion of man and woman or goddess and god. The phallic-like temple at Meru hides a darker sanctuary, a womb-like interior and yonic shrine. In ancient India, male and female was as sacred as the gods, and the lingam and the yoni were specific totems of attraction, part of the ritualized spiritual sensuality associated with Shiva and his Shakti consort.

Dionysus in Greece had a similar following, and among scholars there is a belief that Shiva and Dionysus are inextricably linked. Dance, sensual delight, and sexual pleasure were all revered as sacred art. Dionysus' mythology encompasses the harshness of nature and the beauty of bestiality. Women like the Maenads – wild women huntresses preying on men and beasts – and those who honoured his presence at Eleusis as a symbol of rebirth, are known allies in his erotic regenerative power. He carries a phallic staff and arrives when least expected. He is chaos as Apollo is order. Dance and ecstasy, Dionysian rites of passage, and erotic possession are all forces which, if allowed safe expression, give experience of the sensual world a different meaning. It is exaggeration and abuse of this power that deforms and transforms erotic sensual love into pornographic violence or the 'whips and scorns' of the sado-masochist.

The Middle East saw attraction as a means to an end. The early dances of the Sumerian priestesses evolved into ritual and sacred dances for women's eyes only. Yet in Arabia, the mind of man was conjuring up his own ideology. Extraordinary cosmopolitan cities like Babylon filled up with slaves, musicians, writers, astrologers and dancers from all over the Middle East. The power of the attractive woman became a potent source of inspiration. Slave singers were erotically arousing, their words suggestive and alluring because they spoke in obscure languages. From this cultural *râgout* the Arabs devised two types of love to suit male sexual needs, one of which, ironically, was to be the source of courtly and troubadour love of the Middle Ages. This was known as love-desire, where the woman was alluring, provocative, ravishing and knew how to use her wit, charm and beauty to keep a man drooling on the edge of his seat, or more probably on the end of his camel. Pleasure for the gentleman was in the hunt, for as soon as he had captured and ravished her, he would move on to another likely enchantress. Courtesans apparently enjoyed this game, mostly because they were paid well for each man they seduced. Under Muslim law at the time, a slave could buy her freedom. Liberation of course brought other chains.

China has had changing and fragile relationships with attraction and beauty. The binding of feet became popular around AD 1000, but to keep a woman's foot as tiny as possible is no longer considered a delicacy among sensual delights. The tiny foot was perhaps suggestive of the other end of the leg. Beauty was not seeing the ankle or leg or foot, beauty was in its suggestion. The erotic tiny foot was perhaps also a perversity of man's own need to establish dominance in the bedroom. Tiny feet are like babies' feet, vulnerable, virginal, spotless. The ungrown and the innocent is the unknowable and becomes easily corruptible. Naivety is attractive, and sexually it can be awakened. It may also have been an early form of bondage – the psychological power of unbinding may have been more arousing than was ever dreamed.

*Tear off your footbindings and come into my arms my love,
you will curl your toes with their new-found freedom, you will
curl your toes with the pain of my love.*

Confucianism confused the earlier Taoist spirit of male and female harmony. By the seventeenth century even Chinese writers were disillusioned by love and the senses: 'If in his travels he found a woman worthy of his attention, he looked her full in the face. If she was modest by nature she lowered her eyes and passed him by . . . If she was afflicted with the same weakness as himself she answered his gaze, and between the two of them arose that silent understanding, that "exchange of love letters from the corner of the eye" which, alas, enables such outrageous friendships to be struck up. For a He as well as a She, such a pair of eyes is indeed a disastrous gift of nature.'[7]

So in Confucian Chinese intellectual morality, the eyes are the key. It is the sense of sight which lures men and women into traps of enchantment. Desire is seen as a 'weakness' and the power of sexual attraction a 'disastrous gift of nature'. Of course, as in any other patriarchal society the man must first find the woman 'worthy of his attention' before he would even allow himself to be tempted. Here attraction is first in the eye of the beholder, yet there must be someone upon whom we can cast our eye.

Safer at Night

Maybe this is why mankind, more than any other species, copulates during the night. The enchantress, or woman, is now in darkness. The desert sky will be black and filled with stars. Her eyes and his eyes will be just other stars, disguising emotional feeling, undoing the real darkness. Day-time pleasure is vivid and earthly, the eyes reveal everything. The skin and flesh are lush or monstrous, but always tainted, never as pure as refined expectations. In the harsh light of day his body may seem to him ugly, demanding and potent, hers may seem to her imperfect, flawed and wanting. But at night, the clumsiness of sexual love is hidden. What but the darkest night can hide our fears, our feelings and our senses? Then we discovered the light-bulb, man turned it on at night and he found woman again, chthonic, wild and beautifully chaotic, her eyes lit by his light, now full of desire. Dionysus discovered in the blue flames of desire. Civilization had brought him back to nature with a circle of fire round the water of beauty's charisma. Aphrodite does not let go her hold so easily, she has found us wanting, and given man a light-bulb to shine on woman's femaleness to lure him back into the darkness of chaos and creation, again and again.

Disenchantment

Each person, whether male or female, attracts and repels, exudes their pheromones and desires another. Interaction between male and female creates life, and our sense-perception enables that procreative art to take place. There is something else that is beyond merely bodies, fertility, test-tubes and baby-making, there is more to the sensual than just body senses and descents down the basement steps into the porn shop. The sense of beauty is both one's own beauty, a sense of soul, and something more, something we sense is deeply lacking within each of us.

There seems to be a current collective *Angst*, a kind of disenchantment with our sense of incompleteness, just as the Greeks felt. We know 'intuitively' there may be something missing in our lives. To many of us it may seem to be role-models, parental voices, goals and visions, or simply someone to love. We are no longer spellbound by the film stars and sex goddesses who are paraded before us, because we know they are not immortal. They are human creations. The sex goddess, the idol, the film star, the hero and the explorer, are fallible. They only become immortal when they are dead. As with Diana, Princess of Wales, a vision of someone's loveliness becomes 'beatified' rather than beautified, and we cling to the image as we once clung to the stone skirts of the Virgin Mary. But fallen idols are always fallen. They no longer enchant us. We apparently have nowhere left to project our sense of meaning and to see the divine reflected except in each other. But there is a deeper disenchantment at work. Simply, there is nothing left to enchant us.

It is in relationships that we may find ourselves through pain, longing, desire, sexual grief or sensual pleasure. But our intuition tells us this is not enough, because we only see relationship as a means to an end, and that end is usually marriage. Marriage in itself however, never expressed a capacity for total fulfilment. It is men and women who have twisted the concept of marriage into being a bed for completeness. Marriage to the ancient Greeks and the early Christian Church was always deemed a safe-house, a necessary contract where love, desire and Eros were of little consequence. It was a contract which ensured the continuation of the ancestral line, to seed and to breed. Even today, in many countries and cultures, arranged marriage is the norm and we in the West look shocked and think, how could they? Isn't marriage all about love and romance? Who is kidding who? Marriage can never live up to our heady expectations of it, the white veils as vulnerable and easily torn as they were for Helen and Menelaus. Eternally yours, but with no soul.

The Greeks turned their ideas around and rethought their ideas about women as they rethought man. Marriage was a containment for the forces of nature that women had for so long symbolized. The sensual was seen as belonging to woman, and particularly woman's sexuality. Respected and yet feared, the power of woman's femaleness was always a reminder for the learned, the powerful and the elite of Greco-Roman society of nature's control so that civilization and order could exist. Out of a handful of Greek scholars and poets came the beginnings of misogyny. Women who stepped out of line in marriage, like Clytemnestra, were deemed to be, like all woman, to blame for man's own weakness. When Clytemnestra, 'Killed her wedded lord . . . a song of loathing will be hers among men, to make evil the reputation of womankind, even for those whose acts are virtuous'.[8]

Beauty, the physical senses, awareness and the feeling senses bubbled in a lonely cauldron as the quest for filling the gap between mortal 'me' and the unknown became furiously philosophical rather than deeply felt. So where did feelings and sensual desire hide? We shall remove the veil and find them lying together; we shall look closely at how sensuality became grafted onto sexual love, for they make excellent bed-fellows.

NOTES

1. Keats, p.87
2. cit., Dijkstra, pp.306-7
3. Hubris is a Greek word meaning self-inflation, an outrageous arrogance.
4. cit. Friedrich, p.117
5. Thornton, p. 86
6. Fantham, Foley, p.240
7. Li Yu, p.15
8. *Odyssey*, 24.191-202: cit. Fantham, Foley, p.39

PART 3

Pandora's Jar – Sensual Illusions

'There is the opposite type of woman, who is a great danger to the health and even the very life of her husband. I refer to the hypersensual woman, to the wife with an excessive sexuality'.[1]

CHAPTER SEVEN
The Rustled Sheet

'By the way they adorn themselves they first lead their [men's] minds astray, and by a look they instil the poison, and then in the act itself they take them [men] captive.'[2]

It seems that women and men have always danced precariously on the edge of sexual relationships, sensing both the inherent danger of nature's tidal wave of emotion, and the joy of the purring shingle of first love. Our one-sided perception of sensuality has lassoed it entirely to the world of sexuality. This is our ego-sense attempt to rationalize the sense of mystery, sense of being and sense of romance, as belonging only to physical gratification. The sensual's true value as an erotic interface for body, soul, mankind and nature became hidden under the silk sheets of sexual love. But when did the bed-covers first get rustled?

The Ruin of Mankind

The ancient Greeks are not wholly responsible for the Western split from nature and woman, but they did set a precedent for misogyny and the later misinterpretations of women's nature. Greek writers searching for answers both respected, desired and feared women. Fear and insecurity often lead us to hate. We hate those things in others which we fear in ourselves, particularly our most unknown failings or disowned qualities we are convinced could never be ours. Prime instinctual survival mechanisms are still locked away in our genes or psychic cogs. Instinctive fear is an ally in our survival, but ego consciousness produces a hatch of alien phobias and anxieties, the price we pay for being human.

Misogyny was rife in ancient Greece as misandry is rife now. Equality and liberation has become like the Medusa's head, which many women carry around to turn on or off any man at will. Women and men are in this dance because we have to be. We are each other's terrorists. Many men have as much difficulty in embracing their own feminine image as women do in realizing their own masculine one. When a woman is in an intimate relationship with a man, when she is 'manned', she and he may have a chance to grow, weave, eroticize and love. We cannot be without each other.[3]

The Short Straw

Around 600 BC lived a Greek poet called Hesiod, who could only see through the eyes of his culture, through the eyes of a man, and through the eyes of one who dreams. He could only listen to the world through the deeply ingrained responses of his generation, resonating to his cultural expectations, social genres and family psychology. He could only touch the stones on the mountains with hands that knew woman as his ruin, and he could only taste and smell the morning dew or the fields of wild garlic with a faint recollection of something long ago, something pulling him from heaven, or was it earth? His sense of woman was as remote as his sense of the feminine.

So he wrote about the beginning of time. Sharp, cutting and pungent, Hesiod revealed the story of the gods and goddesses and the earth's beginnings. He knew chaos preceded order, and gods battled with giants, and Gaia, Mother Earth, slept with Ouranos, the God of the Sky. Hesiod's myth is one of the most profound visions of the beginning of creation, although many critics argue that as a literary device his work is indigestible. Sensually repressed, but sexually fuelled, he was simply a man in a world of cultivated men. His words carry symbolic insight, fettered only by personal and cultural fears. So it was at that time women rather than men began to draw the short straws for mankind. Why?

The great thinkers and philosophers of 600 to 300 BC are no different from us now. But a collective split away from nature, and the birth of science versus art, nature versus culture, was the shape of the waves of that time.

Hesiod's other great work was his depiction of the myth of Pandora. For Hesiod, Pandora is the creatrix of all man's woes, desires and destruction. Pandora was created by the gods and moulded into shape out of earth and water. This is a valuable beginning, for already earth and water suggest something more carnally instinctual and intrusive than abstract rhetoric. Pandora does not just represent women; she is the prototype of the dangerously sexual woman. Aphrodite was an ambiguous goddess, Helen powerfully beautiful, but Pandora was the first truly dangerous woman.

Pandora's Real Secret

Briefly, the myth of Pandora's jar (it only later became a box) is about Zeus's revenge on Prometheus for stealing fire from the gods. Once formed from earth and water she was invested with various qualities by the other gods and sent to Epimetheus, Prometheus's rather thick brother. Prometheus means 'foresight' and Epimetheus means 'hindsight'. Epimetheus can only see things in retrospect once the terrible deeds of Pandora's foolishness have been done. Forgetting the warning 'never accept a gift from Zeus', and with no foresight, Epimetheus was overcome by desire for Pandora and immediately took her. Among her god-given attributes were 'charm, and painful strong desire, and body-shattering cares', from Aphrodite. Hermes gave her 'sly manners and the morals of a bitch', and 'lies and persuasive words and cunning ways'. Now the 'deep and total trap was complete'.[4]

Pandora and Epimetheus lived in sexual bliss until one day Pandora could not resist doing what mortals do, and opened the mysterious jar which she had been forbidden to open. Whether fate had simply enabled the jar to be there in the first place, or whether the jar had already been placed there on purpose by the gods, we do not know. Curiosity was another attribute given from Hermes, and of course, Pandora opened the jar. Like Eve, she transgressed, and did exactly what was forbidden. To the Greeks, as later to the Christian Church, this was woman at her most chaotic, unpredictable, dangerous and alluring. To upset the flow, to challenge the weave, to rethread the needle of Fate is one of the most unstable acts that mankind could ever imagine. The opening of Pandora's jar was a metaphor for all that was transforming in the great pagan civilizations. Curiosity, disorder, chaos and decadence, waves and rhythms of life following one after the other and the most profound motif nestling in the image of what man thought he despised and feared most, the feminine. So Pandora opens the jar, filled with all the disasters and ills of mankind. These include volcanic disasters, earthquakes, disease, pestilence, war and even impotence. The only thing she managed to close the lid on was hope.

What Is Hope?

We have a strange relationship to hope. We think it a beneficial quality, yet even the Greeks were unclear about its function. The remaining *solutio* saved in the jar for mankind is actually often a curse rather than our saving. Sexual hope may be redemptive, but it is also reductive. Hope's epithet was *tuphlos* which means blind. Think of Epimetheus who could only see things with hindsight, and the last remaining solace for mankind was to be hope, our blind expectations, particularly in sexual and intimate relationships. Is this a solace or a sense of doubt? If we are always to find answers in retrospect, then what use is hope? Does it not lead us into illusions in the very relationships that we 'hope' will last? Are we not framed by our own expectations?

To the Greeks, and then later to the Christian Church, hope in the marriage union meant the salvation of future generations as carriers of the seed of the father, ensuring the continuation of the dominant male identity. Pandora's jar is then both a dark symbol for woman's womb, carrying evils and 'hopes' of mankind, as well as being a vessel for male potency. Pandora herself is the sensual woman of man's imagination, dangerous and instinctual, beyond his wildest dreams and yet buried deeply in his wildness of desire.

Pandora's secret, however, is the paradoxical nature of hope, for hope is the one quality which rocks the very boat which mankind's arrogance attempts to steer. Hope is the one quality which in itself relies on Eros. Every time there are high expectations, hopes and illusions, we are forced into Eros's domain, a kingdom of transformation unwelcome to most, but inevitable to many. Blind hope leads us into Eros's arms. We must change our perspective, realize our illusions, and cherish the temporal as we do the perennial. Erotic relationships do not stagnate, they may die, or change, but they leave us blessed. Every time hope gazes into the eyes of one who is sexually alluring, or one who sensually knows, hope expects to find herself there. 'She' may do so, but at a high price.

The Trap

Sexual expectations became our sensual traps and snares. Women became collectively the embodiment of men's own dark nature. What could be done to tame the wild woman, her sexual desire no longer in line with the symmetry and idealized beauty of the Parthenon and the growing new world?

Courtesans and Sacred Harlots

The early civilizations embraced their sensual world through worship and ritual; the sacred and the profane were inseparable, and both were honoured. Ancient Mesopotamia and Sumeria propitiated women's sexuality as a part of their spiritual process. The earliest temple prostitutes were considered healers through their sexual secretions. A Sufi proverb suggests, 'There is healing in a woman's vagina'.[5] As their powers were acknowledged, so too was their status. The Babylonian *naditu* were prized for their extraordinary intelligence, their conversation, their profound insight, and later, in classical Greece, the hetaerae were renowned for their graces in the arts. These women were more courtesans than temple priestesses. They were beautiful, witty, wise, vibrant and trained and educated to offer the most skilled of services. Wives on the other hand, were usually treated as subordinate and of lower rank.

The temples of Aphrodite had long been filled with as many as a thousand sacred harlots at any one time. This was big business, but it suited both men and women equally. For a while at least, the power of women's sensual world, both feared and revered, was giving them equal status in some facets of society. The Dance of the Hours, from the Persian word *houri* and Greek *horae*, was performed by temple priestesses at a specific hour each night to protect and honour the sun-god Ra's boat as it crossed over the dangerous chasm of the underworld. This later became a pagan ceremony of the divine *Horae*, and the oldest Hebrew folk-dance is called a *hora*.

Geishas

The geisha girl has generated a stereotype of the perfect Eastern woman in the Western mind. Geishas, however, were never sensual seductresses. Western men have an image of the geisha as the total woman, obliging, feminine, contained, wifely, industrious, clean, painted and sexually submissive. Tokyo was originally called Edo. Even in the mid-eighteenth century pleasure quarter, the geisha's role was purely one of musician, singer and dancer. The geishas were strictly trained in the arts of the muses. They rarely turned to prostitution for fear of political reprisal, and the Western image of the erotic and refined courtesan all rolled into one impeccable image of woman is an illusion and assumption. Japan is a hotch-potch of cultures. It advocates pleasure and seeks to please without value judgement. It is both permissive and yet restrains eroticism. The difference in the East is that much of Japan's culture is an import from Chinese and Indian societies and philosophies. Sexuality in the East was an ancient sacred art, and the harmony between men and women, although subject to the changing dynasties, wars and religions, was honoured rather than despised. Yin-yang and the *Kama Sutra* made sexual mincemeat of the Victorian sex manual.

Throughout ancient eastern religions and mysticism, the lake of the waters of birth was known as the source of creation. Water is a feminine element in Eastern philosophy, thus to dive into the waters and be initiated into the sexual act was how a man could achieve spiritual enlightenment called *horasis*. However the New Testament (Acts 2:17) preferred to translate this moment as 'visions' rather than sexual release. It seemed that the 'dangerous woman' could play a valuable part in man's spiritual and profane lifestyle, as long as she remained contained. Even Hesiod managed to admit that the sacred whores of Aphrodite 'mellowed the flavour of men'.

Thargelia and Others

With the power of the hetaerae in Greece came new fears for men. These were women who were not only beautiful in body, alluring and provocative, but also had wit, wisdom, and most worrying of all a desire for power. One of the earliest courtesans recorded was Thargelia, who acted as a secret agent for Cyrus the Great of Persia in the sixth century BC. Throughout history, prostitution, courtesans and wild women are often linked to the art of spying, deception and intrigue. Thargelia was noted for her art of seducing men into telling her their political secrets and then ensuring they were rewarded with easy victories over their competitors. Her distinguished lovers eventually offered to give up Ionia peacefully to Cyrus's domination.

What was her power if not seduction, sexual provocation, deeply erotic love as well as feeling? Whether it was feeling for Cyrus the Great, her love of power and persuasion, or the love of her work, is irrelevant. Thargelia was living out her identity as a secret agent/courtesan. This was her survival, this was her gambit in the world of men. If civilized power meant you were less vulnerable, then women could discover this too. PMT and menstruating charm were no assets, but an intuitive sense was. By developing the sensual arts, and particularly those which seemed to draw the other sex irrevocably to you, the feminine qualities were being made manifest. Other women also rose to high status or gained power through their noble art. Thaïs of Athens was the renowned mistress of Alexander the Great, responsible for the burning of Persepolis; she later married Ptolemy I and became Queen of Egypt. Aspasia ran a literary and political salon in Athens and was known for her part in the Athenian war on Samos. Like the goddesses that preceded them, these Greek women were influential, equal in status to their men.

Theodora

The dancer who became Empress of the Eastern Roman Empire would not shock the world now. We are primed for exactly that dream, deprived kid becomes movie star or marries millionaire, prostitute is pulled off the streets by wealthy business man; and some women seduce, manipulate and sleep their way to the top, driven by the desire for power and success, no different from any man. The only difference now is that women can be achievers without being married or someone's mistress, although they often are both.

Theodora was a dancer, comedienne, actress and courtesan in the sixth century AD. A rich client took her to live in Egypt, then she returned to Constantinople alone. There, her status assured, she mingled with politicians, church leaders, religious thinkers and the heir to the throne, Justinian. The forty-year-old, unmarried emperor to be fell passionately in love with Theodora, and amended the laws so that he could marry her. During the revolts of 532 BC she persuaded Justinian to remain, rather than run. In a powerful speech in front of the senate she said: 'if I had no safety left but in flight, I still would not flee . . . I shall stay. I love the proverb: Purple is the best winding sheet'.[6]

Eros's cloak is purple, it winds around those that cannot escape the pull of the erotic connection. There was nothing Justinian could do to persuade Theodora to leave. He was tied inextricably by his erotic desire for this woman, however much she may have been reviled as a sensual beast, a sexual vampire, a torrid and power-seeking woman. It was through Justinian's eyes, ears, touches, caresses, his senses, that very unbinding feeling from his soul that he could only respond to her. He was bound in her winding sheet of erotic love. Theodora courted her own sexual power and Justinian, like many men, painfully ensnared in his carnal melancholy, does not realize the sensual huntress that he perceived may also be a reflection of his own hunted shadow.

A mosaic portrait of Theodora, resplendent in her jewels, a benign but unfeeling face revealing the cautious power of her Empress beauty, hangs in a church at San Vitale, Ravenna, Italy. Theodora became a devout Christian, yet the image of a sexually experienced, provocative woman is strangely out of place in the Christian Church that was to condemn prostitution, sexuality and the nocturnal, beautiful, moonlit woman that Theodora was.

Sensual Desire

'To desire and to see through desire, this is the courage that the heart requires.'[7]

Desire comes from the Latin word *desiderare* which literally means 'absence of stars' or 'to miss a star which is looked for'. Also, 'to expect from the stars'. When we have no stars to guide us, in other words no map to follow and no sense of rational constructed images, we may become lost. Without the stars, particularly the one we hope for or 'expect', we become subject to the energies of the unconscious. There nature and her chthonic earthly passions can take us over, fill us with terrifying wanting, urge us into fantasies and illusions about another. We are trapped in the feeling, it is ecstatic, unworldly, yet at the same time dangerous, foreboding and infernal. This is when Aphrodite and her envoy Eros come to fill us with desire. Desire quickly captures all the senses. It feeds on all our senses, setting snares, surprising us with quicksands of obsession or the shock of perfect bliss. Without sexuality, desire would be a force so intense we could not release its binding energy in any other mode. It holds us wrapped in its arms, touches us with frozen fingernails or the scalding irons of Agni, the fire god. Desire, like Eros in Greek mythology or Kama in Indian myth, forces us into a state of the irrational. Desire is the natural chaos of missing our star, missing our way and being led astray by the blackness of the gods and goddesses of the night, or by our 'expectations' and our blindness to the truth.

For most of us, desire walks hand in hand with us, whatever the moment. It is when we desire another person with high expectations and hopes that we may find desire brings first a temporary wish fulfilment, then finally disillusionment. For we place desire on the other and see what we 'want' or yearn for. This is not just sexual desire, it the desire of total rapture, of pure ecstatic union with the image one has imagined. But desire feeds only on that which it cannot have or be. For once we become familiar, or no longer want that very quality we have imagined the other to be, then desire dies, until it or the other becomes unattainable again. The veil is lifted from our illusions only when we take back the image and claim it as our own, but we seem to prefer living with, rather than without, the illusions. Strangely, many men think it is only they who are tormented by desire, recklessly and hopelessly torn by woman. This is a collective perception, but it is understandable considering how the past several thousand years of humanity has played with Aphrodite's toys of passion. Born from the bloody foam where Ouranos's genitals fell in the sea, she sends us 'the sweet joys of seduction and sex, the loveliness of desire' to fool us into love. Once there we are caught *in flagrante delicto*, and like Zeus's trick with Pandora, 'the deep and total trap' is now complete. Desire is sent from the gods whether we like it or not. What we do with it is another matter.

Desire is our way in to discovering our sensual world, but sadly it is rarely an entrance graced with sweeping gowns, gentle kisses and delicate lips. It is not an entrance across a golden threshold but an open door to self-confrontation and initiation into the mystery. Without desire, we would never move erotic energy in the way that we must. Without desire which evokes the senses, men would not love women and women would not love men. Desire breeds love and demands love. Yet there is always a gap into which we fall when desire moves out of our arms as unexpectedly as when it arrives. Sensuality and desire are partners in the same divine *crime passionel*. To be filled with desire is to be filled with the gods. For human beings the manifestation of this feeling of possession is a craving for sensual and sexual ecstasy, but if the gods were left to their own devices it would not be sexual union that filled the gap, it would be holiness. James Hillman writes, 'Desire is holy, as D.H. Lawrence, the romantics, and the Neoplatonists insisted, because it touches and moves the soul. Reflection is never enough'.[8] It is only man and woman who have the greatest difficulty dealing in the language and passion of the gods.

> *Night falls and the darkness of desire lingers like the taste of her lips on mine. Senses curve my mind into turmoil! No thinking know, only wretched fantasies, imagined completion.*

NOTES

1. Dijkstra, p.334
2. Quoted Baring and Cashford, p.513
3. Homosexuality is here not entering the argument because this book is basically concerned with heterosexual relationship.
4. Hesiod, p.61
5. Edwardes, p.96
6. cit. Lindsay, pp.294-5
7. Hillman, p.15
8. Hillman, p.273

Chapter Eight
The Dark Feminine

'There was a time when you were not a slave, remember that . . . you say there are no words to describe it, you say it does not exist. But remember. Make an effort to remember. Or, failing that, invent.'[1]

We need to understand what is meant by the 'dark feminine' image and how it is a dilemma for both man and woman in the world of sensual relationships. Our senses enable us to reach out towards another, and also to draw another to us. The sensual is sensed as a yin or feminine quality, just as the darkness is.

The night before the new moon is the darkest night of all. We have instinctively learnt to be wary of this lunar, haunting darkness, for it is symbolic of danger and death. It is the time for Hecate to roam, for vampires, ghosts and earth mysteries, for pagan love and screeching owls, those objects and nature's creatures who remind us of our own nightmares. This moonlit danger is also seen as feminine. Thus woman not only embodies fate, but she also embodies the darkness. If our instinct tells us to watch out for danger, then our egoic sense feeds on this natural awareness and turns it into conscious fear. Women cannot be separated from the image of darkness when the mind has interwoven the two. Our left-handed, mysterious enchantment is malevolent to those who relinquish nature at the expense of culture. Historically, we have assigned the masculine as solar, or the solar as masculine. We have also assigned the lunar as feminine or the feminine as lunar, and the shadows cast by the moon are those which one cannot bear to see within oneself. These shadows are the sensual baseness of all that is considered feminine or yin, and that includes woman.

What Are Our Shadows?

Our psychological shadows contain those qualities, whether good or bad, which we disown, or make us feel inadequate. We push the parts of our personality we don't like well and truly behind us, and like our physical shadows they follow us around, shocking us at times when we turn to gaze at them, or they appear as a reflected image in the mirror, an unpredictable fascination with ourselves in a variety of disguises. The sun's shadows are usually considered similar to the qualities we don't know we have and which we may envy or admire in others. But the moon's shadows are shadows of silver-reflected light. Sombre and tenebrous they are the shades of obscurity, of *incognita* veiled by her darkling angels.

These shadows are the qualities in oneself that are fearful, vile, 'not mine', disowned and despicable. We have 'no desire' to witness our dark shadows, for they mean we may have to confront the sense of darkness within ourself. But the ego's will for no desire does not dispose of 'desire' so easily. Desire leads us into places and into people where we may find our own shadows suddenly getting in the way of relationships. Desire keeps us vibrant, alive, keeps us aware of the chemistry or the mystery which flows through romantic and sexual love. Here our shadows may manifest themselves in the form of illusions or expectations, as manipulation, arrogance, rejection, self-sabotage and pretence. Desire is not of ourselves, it seems to come from the unknown, and brings with it those darkling angels, so they may confront us when we least expect them.

The 'feminine' or darkness, or receptivity, or lunar light, has been gradually devalued over several thousand years simply because as humanity leaned more towards science, civilization and conformity, it inevitably moved away from the irrational, and subsequently the sensual world. The Serpent Goddess and her secrets have been banished from the collective, just as Eve was carefully written out of the Paradise script.

Expressing our passion or our pleasure does not come easily in the West. We have conditioned ourselves to reject, disown or deny those qualities which our cultural construct finds undesirable. We stamp out passion, sorcery, the unknown, mystery and pagan gods because for the past two thousand years we have been taught to believe in one god. Desire often brings with it the undesirable. We have another shadow behind us in the form of the Judeo-Christian Church. It is not easy to live with such an ominous shadow, because we have our own individual shades to contend with too. The abbeys, cathedrals and churches of solar wealth often cast long golden shadows across our conscience.

The early church leaders feared the quality of sensuality because it was antithetical to all their beliefs about transcending the animal within. Slowly, as misogyny resurfaced, the more corrupt members of the Judeo-Christian Church fed the body of mankind with a festering poison through a numinous drip. False images of the feminine as entirely dark, entirely lit only by the umbras of hell, forced mesmeric illusions into the collective soul-mind. Woman, for all her enchantment, became disenchanted and she too began to believe that she was to blame for man's disasters and destructiveness. She became simply the personification of the dark feminine.

In Western history the image of all that was sensual became intrinsic to the sexual, just as Eve absorbed the feminine darkness. If patriarchal scribes, priests and church politicians could project their own shadow onto one scapegoat, then it had to be in the best piece of fiction writing ever, the Biblical story of Adam and Eve. But then, as some men do, they forgot about Lilith.

Lilith and Eve

These archetypal women are two of the most powerful images of the dark feminine embedded in our Western psyche. Lilith represents everything in woman which is spell-binding, enchanting, rebellious, untameable. She would say, 'No one is to blame'. Eve embodies everything in woman that is child-bound. She is man's ideal model of duty, obligation, obedience and wife. She would say, 'I'm to blame'. Lilith is wild woman sexuality, spirited, nature's inferno; she embodies men's fear of their own darkness. Eve does too, but as the subjugated woman, dominated by the power of masculinity, men's own defensive.

Lilith

Historically, Lilith has not had a good press. Yet we need to make her apparent darkness conscious. She represents the bewitching sensual awareness which one needs to express and acknowledge in oneself. Lilith personifies all that seems taboo, forbidden and dangerously beautiful in woman. Eve, at least the Judeo-Christian Biblical image of Eve as Adam's consort, embodies all that man apparently ever wanted in a woman – submission.

Lilith's early elusive life began in the pleasure domes of Sumeria. As Inanna's handmaiden, it was Lilith's job to gather exquisite young men from the streets and bring them to the sacred temple, where they would take pleasure in sacred sexuality. Her epithet was 'beautiful maiden'. There are many versions of her early myth, but it seems likely that with the incoming power of patriarchy, Lilith had to be assimilated. The sexual power of woman had to be redirected, sensuality could not be the answer to the divine, and subsequently Lilith became identified as Adam's first wife.

Probably around the eighth century BC Lilith was merged with an older demon, a child-killing witch called Lamshtu, although in a later eleventh-century Cabbalistic work, Adam and Lilith are an androgynous being with equal rights. Early Biblical stories describe her as Adam's first wife, troublesome and wild. Refusing to lie in the missionary position, she preferred to take a dominant sexual role and sit astride him. In their conjugal dispute she eventually left him, flew off into the air and went to live by the Red Sea with lascivious demons as her consorts. She gave birth to innumerable 'Lilum', demonic babies who were later to seize men in the night and have sex with them while they slept. Lilith became distorted into a she-demon, with an insatiable lust. She would engage with men who slept alone, causing erotic dreams and nocturnal emissions.

Lilith was both desirable and dangerous, her image a complete sensual illusion. Virginal in the truest sense of the word, belonging to no man, self-possessed and autoerotic, she was the first *femme fatale*, seductive and beautiful yet viscerally adept at throttling her men. Integrating this image of her as a kindred spirit was impossible, yet desirability and danger are ferocious siblings.

Apart from Adam's first wife, Lilith has been called a screeching night Hag, a seductive succubus, demon of the storms, blood-sucker, a night monster, and a wanderer between heaven and earth. She has been described as having a mouth 'set like a narrow door, comely in its decor', and her tongue, 'sharp like a sword'.[2] The Victorian romantic poets were fascinated by her power. Rossetti wrote of her:

'. . . as that youth's eyes burned at thine, so went
Thy spell through him, and left his straight neck bent
And round his heart one strangling golden hair.'[3]

Beast-like woman are also renowned for their ability to strangle. The Sphinx, also known as the Throttler, who lived high above Thebes on a cliff-top would ask a riddle to all who entered the city. If they did not answer it correctly they would be throttled to death. Lilith as an erotic image was often used in literature and pornography to demonstrate the power of woman to strangle her prey during sex. Nineteenth-century artists and poets found her deadly and irresistible, the heart of sensual woman was theirs to manipulate. Twentieth-century 'Goddess' feminists on the other hand see Lilith as representative of liberated sexuality in women.

Wet Dreams

It was said that every time a pious priest or Christian had a wet dream, Lilith laughed. She was the succubus who haunted many innocent and guilty men, stealing away their life-blood, essence and potency, draining them of life as she took her pleasure. Now pleasure, sensual pleasure becomes twisted with images of death, with the most fearful dream of men, to be powerless and lifeless. The gut survival instinct is threatened by the image she evokes. Once a man had experienced the taste of a daughter of Lilith, no mortal woman could ever satisfy his lust again.

But What about Eve?

The myth of Adam and Eve existed a thousand years before Christian leaders corrupted it to serve their own purpose. The Bible treated the myth with scorn, blaming women for the writers' own dark motivations. This Biblical story has had a profound effect on our attitude towards love and sexuality. Like a poisonous wound, it has eaten so deeply into our Western skin that it is hard to find an antidote.

An early Gnostic story, however, suggests a doctrine of equality: that Adam and Eve were both hermaphrodites, and the creative force which enabled them to unite was a feminine principle. Adam was created with no soul, and then it was Eve who breathed soul into him. In the Gnostic scriptures Adam and Eve represent nature separated from the divine, their quest being to find each other and to unite in the divine moment. Gnostic texts also refer to Eve as the 'mother of all living', who created Adam to be her consort. The serpent *was* Eve, mother of both, and may be rooted in the Serpent Goddess herself.

The biblical interpretations created Eve as a less troublesome wife for Adam than Lilith. The Bible plays heavily on Eve's temptation by the serpent to eat the fruit from the tree of knowledge. God throws them out of the garden and curses them for eternity. Eve then becomes an easy pawn in the Church's own dogma, for Lilith's wildness means as a figure of subordination she would certainly have rebelled. But Eve does not. So it is Eve who is blamed for the fall of mankind, and thus Christianity can save 'man' by finding a redeemer. 'Take the snake, the fruit-tree and the woman from the tableau, and we have no fall, no inferno, no ever-lasting punishment – hence no need of a saviour.'[4]

Nocturne

Other dark feminine images have included the Chimaera, the Hydra, owls, bats, the night, Hags, sirens, she-beasts, and an assortment of succubi and terrifying female demons. But one of the most potent images of our dark nightmares we still carry today is that of the vampire.

Vampires live in hope, with high ideals of that first taste of blood at sunset. Here, day-time sleep is a dream-spinner of ancient memories, of the first blood-lust, blood rituals, menstruation and woman's power. Rising as a nocturnal creature to drain the blood of its victim, the vampire lives in the hope of sexual pleasure as well as blood nourishment.

We recognize vampires from their Slavic origin as the undead, corpses returned to drain the blood of the living. But ever since the ancient Greeks believed the undead could be recalled from the underworld with blood offerings, the Western psyche has had a fixed notion that blood could restore the dead to life. The Greek word for vampire, *sacromenos*, meant flesh made by the moon, and throughout European history a popular association has been made between the rising of vampires, the dead, the moon and women.

The first vampires in literature in nineteeth-century Europe were women, sharing the night-time duality of the moon's favour. The vampire myth was possibly reinvented because of the growing fear among intellectuals over the resurgence of the sexually powerful woman. They remembered, unconsciously or not, the ancient rites associated with blood and the moon. Artists and writers struggled with the image of the vampire and their perception of the devouring nature of women. Misogyny was as rife then as it was in ancient Greece, only this time, it was a precocious defence against the emerging new woman rising out of her coffin to attend to her wild side. Many women were ready to reclaim the virago and bite the neck of patriarchal idealism that had swamped the collective psyche for over two thousand years.

'It is to her the name vampire can be applied in its literal sense. Just as the vampire sucks the blood of its victim in their sleep while they are alive, so does the woman vampire suck the life and exhaust the vitality of her male partner – or victim.'[5] The new sexually alert woman was to be staked through the heart before she returned to the safety of the night, and the coffin. This generation of Empire-builders saw the vampire as a symbol of the eternal dark beast which longs to return to woman's body, in constant struggle and agony as it writhes against mankind's civilizing of it. Tame the beast, for it returns unseen, unwatched, in woman's tender soul, 'Whose lust of blood has blanched her chill veins white, Veins fed with moonlight over dead men's tombs'.[6]

The vampire came to represent the feral, visceral, earthly chaos of woman's body, remnants of the ancient past of nature's darkness that man still could not face. Her body still enchanting yet blood-draining.

The most stereotypical vampire was ironically to be a man. Bram Stoker's Dracula quickly learned to feed on the blood of young women to remain young and eternal. He knew the food of eternity, the chaotic inferno of time, had its source in woman, and in her blood. Stoker however, lived at a time when he was confronted by the beast in himself reflected in woman. The two victims of Dracula are analogous to the two faces of woman that were currently in vogue.

Lucy, the wild woman, sexually flushed with desire, was permanently alluring to man, a virago, a nymphomaniac. Her own blood contaminated by Dracula's, she is a demonic vampire. The ultimate turn-on for Victorian man, she was equally lustful, but bestially demanding. Finally, when faced with the crucifix ,'her beautiful colour became livid, the eyes seemed to throw out sparks of hell-fire, the brows were wrinkled as though the folds of flesh were the coils of Medusa's snakes.'[7] The sexual imagery and symbolism is flagrant, but the crucifix became the potent force that could save man from nature's fury. Christianity believed it could do anything with the crucifix, but then they were relying on an ancient pagan symbol, and one which ironically was of phallic significance. In ancient Egypt, the cross was a male symbol of the phallic tree of life, as the circle was a symbol of a woman's vulva.

Could Bram Stoker be faced with something more terrifying than a vampire – was it perhaps the rediscovered power of the demonic woman which unconsciously aroused his pen as he wrote? Once disposed of by stake and crucifix a vampire woman was safe, she could become Victorian man's ideal again. 'There in the coffin lay no longer the foul thing that we had so dreaded.'[8] By the end of the nineteenth century belief in the ideal woman was very much as an image of a sexually 'dead' woman.

On the other hand Stoker's heroine, thoroughly modern Mina, is a civilized woman, a typist and pragmatist, every man's dream wife. She is nurse, busy, bustling wifely know-how in her gowns. She 'thinks' rather than feels, she is the household whizz-kid of the new intellectual progressive shift of the *fin-de-siècle* – a new woman without the undercurrents of sexual stigmata and only the flounces of the *belle époque*. Feminism was curling coils through sensual traps, but man's ideal of woman had all but freed itself from her nature. This new 'She' was to become built for marriage, work and the machinations of civilization. There was a growing sense of something apocalyptic, the First World War already knocking nails in future coffins. Women's role would be drastically changed for the first time in modern history.

Dracula himself, a symptom of woman's apparent darkness, could only be silenced through the progress of society and the daylight doings of men. Those nocturnal memories still jangling in men's pockets as they paid for prostitutes were pushed deep down in the hem of their coats. Circe and Hecate, witches and demons were to become vampires preying on the minds of lust-less men. The new emerging woman was symbolic of the Lilith of man's wet-dreams, first the initiator and then the throttler of his arousal. An eternal problem since Gaia persuaded her son Kronos to castrate his father Ouranos.

Ouranos, the god of the heavens, was seemingly unable to diminish his eternal lusting. With such high expectations of his offspring, his ideals so abstract from Gaia's earthly reality, the beings he created all became monsters and ugly creatures in his mind. Too much for his heavenly idealism, he hid them inside Gaia's belly so he could never be disillusioned. Yet Kronos, in fear of his own possible suffocation, castrated Ouranos at his mother's direction. Aphrodite was born from the foam where his genitals fell in the sea, and Ouranos no longer had a role to play in the creation mythology. Is this what we perceive some men to fear truly? Not so much castration of their generative organ, but castration of their generative idealism.

Vampires are a superb symbol for men of their sensual terror, a neat myth in which women's apparent demonic lusting feeds off every innocent male she meets. The archetypal woman has at many times in history become the ultimate personification of bestiality, chaos and darkness. In the past thirty years the genre of vampire films may also be another statement, an unconscious collective retaliation against the stirring vitality of women, now able to promote the more extreme outlaws of macho-feminism.

But the sense of darkness is the most terrifying sense, which can manifest in both men and women. The dark shadowy feminine hides under many blossom boughs, in women or men who love too much, or hate too much, who betray, or who seek revenge; in men who reject and sexually manipulate, in nagging wives, in whining exes, and in promiscuity and overt sexuality or pornography.

Whatever we disown we may devise in others. Our demons may be cast unconsciously into another, and our shadows then come back to haunt us in real life in real people, drawn to play out those things we reject, deny or find threatening in ourselves. False images may manifest themselves as jealousy, spite, obsession, possession, greed and exploitation. Awareness of the feminine or yin-sense means we may at least begin to own and address our demons, whether they are cast out by the light of the sun, or thrown behind us in the light of the moon.

'Bound Is the Bewitching Lilith'

Lilith became bound, sealed in a vampire's casket, the blood wiped from her lips and a stake run through her heart. But her shadow lives within us still and the sensual darkness is attached to her legacy. She has kissed us all with a vampire's kiss, even though the incisors sinking deeply into our lily-white necks were first polished with Christianity's toothpaste. Lilith's own neck was kissed by those pious churchmen as they thirsted mostly on the blood of woman, but it was not in sexual expectation, not for desire and carnal knowledge – they say it was to redeem her. The vampire has no reflection. Have you ever seen a mirror in a church?

It was through a collective demand that men received the version of the women they desire, and throughout Western history many women suppressed and denied their sensual world for it gelled too easily with the symbol of Lilith's 'blood-red, kiss-provoking lips'. Woman either exaggerated in herself what was being called for, or she denied it. The sensual and the image of woman were becoming intrinsically bound to one another in the collective psyche.

Blame

The sexually powerful woman was to become the dark feminine, she who can take a man away from the subordinate wife and his sense of defensive domination. 'You blame me, don't you', is an underlying message in the bed-chamber, still heard in the shadowlands of suburbia. The urban semi throws out evening shadows longer than a cathedral.

So when this wife, for example a fictitious Geraldine, asks, 'You blame me, don't you?' because he's impotent, he answers, inwardly at first to himself, yes. Outwardly he will of course deny Geraldine is to blame, and reach out for the mug of cocoa and blame himself. But because of his inherited values of the past two and half thousand years or so of woman as split into lover or wife he believes, deep down, in some place he finds uncomfortable, that she really *is* to blame. He knows about Lilith as the beast lover, he has tasted her once or twice: he's found and met Lilith in many women. Even the dangerous contradiction he first adored in Geraldine when they kissed at college. Wow, she was Lilith then, erotic, powerful, seductive and dancing, alluring, teasing, coaxing! But now, dangerous woman has become simple wifey, the home-loving, career-frenetic, manicured nails and microwave new woman. Danger is what seduced him into potency, seduction, mystery, intense desire and passion, not routine and familiar flesh. What is he to say to himself except, yes, I blame you? For she has become Eve now, not Lilith.

The darkness and the feminine are uncomfortable bedfellows in the Western psyche. Lilith is wanted, yet she is wanted only by induction. Geraldine's husband may look for Lilith elsewhere or, like the celibate monks in medieval Europe, he may sleep with his hands clasping his genitals in fear, the crucifix clutched between his phallus and fingers in a vain attempt to keep the thought of Lilith from contaminating his mind, let alone his body. Could it be that Geraldine will resign herself to enjoying the pleasure of his crucifix rather than his loins?

> 'For in your body is the inevitable
> Sting of the Serpent made of the snake's desire.
> The desire he had of Lilith'.[9]

So Geraldine and her husband may have little choice but to sleep restlessly in the belief that she is to blame for his flagging phallus, and he will be saved from his impotence by either a visit from another Lilith, or God. We wonder which he would chose?

> *She is richly warm this one, her perfumed bed my night's desire. If I could take her now, if I could engulf her dark passion and bind her to me. Come Lilith, my body's enchantress, tempt me with your mystery.*

NOTES

1. Wittig, p.89
2. Koltuv, p.40
3. Rossetti
4. Daly, p.69
5. Djikstra, p.203
6. Symons, p.203
7. Stoker, p.218
8. ibid, p.222
9. Symons, p.301

Chapter Nine
'Fair is foul and foul is fair'

'In the self-same point where the soul is made sensual, in the self-same point is the city of God ordained from without beginning'. Julian of Norwich, 1343-1415

Mystery

The sensual is concerned with mystery. All our physical senses, taste, touch, smell, sight and sound, are instinctive and scientifically explained and validated through measurement and logic. But what of our sense of being? What of our sense of love, or feeling, or sense of mystery itself? This sensual 'incognita' is the interface between our known sense-perception of reality, and the unknown. But what is mystery, and why is it necessary to value it?

Ancient peoples gave great value to mystery and the unknowable. Mystery derives from an ancient word meaning to close the eyes or lips. But in our quest for civilization and improvement, we have gradually conditioned ourselves to place value only in the known. We have devalued the mysteries of nature and the mysteriousness of our pagan inheritance. Anything unknown must become known, it has in our minds to be clear and defined; it must have no hiding place; it must be cleansed of enigmatic lures and our eyes must be always open. We acknowledge the sensual when it manifests itself through our physical senses, sight, sound, touch, taste and smell, but when we try to get a 'sense' of feeling, of imagination, of intuition, most of us prefer to turn a blind eye, simply because confronting incognisance is about confronting oneself.

Woman's body was one of the earliest mysteries known to mankind, her sexual beauty and body seductively entwined in a man's arms for thousands of years before he knew her anatomically. His fear and desire caused delusions about her body's power, and respect for it too. It was only when the body became rationalized, that humanity turned to the mind and feelings as a mystery embodied in woman. As a woman was the most obvious image of the feminine principle, it was not so much *femaleness* which was the target, although it seemed as if she were, but the yin-sense of Everyman projected onto the very image he was attempting to define.

Today a woman's body is no longer mysterious in a scientific sense. We know every nook and cranny of womb, fallopian tubes, breasts, glands and ovaries; we know her cycles and rhythms, her ageing process and her youth as we do with men's bodies. Yet somehow there is an essence, a notion, a charm, an elegance and subtle imbuement in a woman's body that can never quite be grasped and made tangible. We still perceive some mysterious quality about a woman that is indefinable, possibly because her associations for thousands of years have been arcane, night-time, dark and devious. Or it may simply be the secret place, her womb, her cyclic power, her biological necessity. Whatever is indefinable about a woman's body it has grace, and it 'closes our eyes'. We want it to be mysterious, so it is.

The 'unknowable' has always been a dilemma for the rational mind. Philosophers, spiritual and psychological thinkers and artists have all battled with the demons and gods which make human nature so rich. Mystery finds gaps in society; like love, like Eros, like soul, it pushes its way into the fabric of our perceived world to entice us away to a deeper place. The unknowable has also been distorted by individuals and cultures as a vehicle for power, greed or manipulation. Pure mystery is not destructive or evil, it is those who exploit and abuse the 'closed eye' who inject power, fear, and scepticism into its very lens. Yet perhaps the biggest and best known of 'incognita's' twists arose through the Judeo-Christian fear of woman being the most flamboyant of mystery's symbols.

Eco-relating

Our feeling world has always given us trouble. Nowadays we find it hard to communicate to others our need for space to 'feel' our way. Because many woman have chosen equality, career, or motherhood, or have a vision of success, wealth and power, they often deny their femininity. In other words they are not connecting to the mystery within; they are not listening to their intuition, their sixth sense; nor are they making conscious their sensual erotic interaction with life in all its facets, including sexual relationships. The reverse is also legitimate. To attach oneself exclusively to a feminine image means a woman may become overtly promiscuous and attached to Aphrodite; or attach herself to spirituality and never touch the ground; or be too bound in the role of mother and attached to a lunar image. Similarly, a man may deny there is any mystery in his life and attach himself totally to all that he perceives as real, i.e., materialism, work, power. Or he may bind himself to women or something labelled 'mysticism' as a source of enlightenment, fearful of the 'real' world and its hierarchical order and system.

Becoming parthenogenetic, self-procreating, could be a reality for woman in the future. She may not need her man for much longer, apart from his seed. This is frightening to men, but it may also mean we tip the ecological balance of nature too far. What do we do with the men we create, or do we eliminate the male factor all together? How will women enjoy sexual pleasure? Always with other women, or alone? Or will a man become a woman's accessory, her adornment or sex toy, her transitional object? And will sexual pleasure be irrelevant anyway?

Eco-relating is to be in relationship to all things through one's sensual awareness of others and of self. If we are in good relationship with each other, then maybe the earth can benefit from the radiance of pleasure. To have a sense of mystery is to live with what the Zen Buddhists would call 'the sound of one hand clapping', what the medieval alchemists would describe as 'as above so below'. To have a sense of 'incognita's' soul is to read the words of John Donne, and follow his advice:

> 'Goe, and catche a falling starre,
> Get with child a mandrake roote,
> Tell me, where all past yeares are,
> Or who cleft the Divels foot.'[1]

Although partnerships are the most important thing in most people's lives, we are beginning to find other ways of reaching towards an erotic inter-connection. We are able to be more creative with ourselves, and subsequently our relationships have a chance of improving for the benefit of us all. Eco-relating is to have a sense of love's movement. It is to be in physical contact, sexual contact, erotic contact, feeling contact, imagined contact, mind contact with life and nature. It is to be open to 'incognita' and her imagination. It is to be in-sensed.

We no longer have a powerful archetypal figure or icon of mystery to reflect upon, apart from religious mystery. We have tried in this century to place value on movie stars, and glamorous men and women instead of saints and sinners. We have even tried to bring the Great Goddess alive, perhaps as a necessary balance to the overt masculine principle we have worshipped for a thousand years. Whether its the Marilyn Monroes, the Valentinos, the Grace Kellys or Dianas, the mystery that is theirs is also ours. Aphrodite is attempting to make a comeback, her vengeful, jealous nature seeks always to arouse and to eroticize. She has a shifting nature, and one that has ridden the storms of mankind's obsession with yang power, just as men and women have had to do for the past several thousand years. She is also a force to be venerated again, simply because if our relationships are in danger of falling apart we need to know how to be creative with the loss and the pain. The most mysterious thing about oneself is oneself. As Jung wrote, 'what is closest to us is the very thing we know least about, although it seems to be what we know best of all'.[2] Through participation in relationship to the world, the sense of the unknowable may begin to be accepted again.

We are, it seems, becoming global people, rather than men and women separated by one-sided dogma and cultural constraints. Yet essentially the rift is still so deep that woman and all 'feminine' images may always be a mystery to men. Sending depth sounders down in an attempt to measure mystery which is measureless is as absurd as a square circle.

Shakespeare has a lot to say about the mysteriousness of nature. Like many of his characters, Hamlet also has a lot to say about Shakespeare, probably one of the most 'sensual' of men, given the cultural constructs at the time. Shakespeare, or rather Hamlet, gently reminds Horatio, 'There are more things in heaven and earth, Horatio, than are dreamed of in your philosophy'. He really should have said this to Laertes, as we know, yet the profound truth is that mystery lives through every manifestation of life, as it does in every star in the sky.

So if the sense of the sensual is intrinsically a mystery, why do women attempt to hide it, disguise it, or suppress it? Why is it that some women are vain enough to watch themselves in the mirror, but are unwilling to express themselves as sensual people? What is it about the interface of the divine and the profane in us that we disown, or disavow? Perhaps it is because both mystery and sensuality have been split off from linear thinking and are contradictory to the image of one god. The paradox is that mystery has been stolen from the body and all things pagan and given to the mind. 'Incognita' has become masculinized into incognito, and mystery swings ominously like a knife-edged pendulum as belonging solely to a monotheistic religion.

Héloise

Christianity enchanted mankind with its saviour and its message of love and faith. It is not Christianity itself which was any worse or better than what had been before, but it was the corruption of its voice and belief that forged a deeper wound in the heart of the feminine principle.

The abbesses and female scholars of the early Middle Ages were presumed at least to have souls, previously only believed to reside in men. The ninth-century woman of pious and learned faith held an unusual equality in Europe. A growing sense of mankind's own extraordinary awakening consciousness and civilizing power alerted the churchmen however to the underlying urges of their own sex and that of women. Menstruating brides of Christ might soil the holy places, and so women were excluded from mass. Woman was more likely to disturb and tempt devout males by the very presence of her body, double messages of sexual persuasion and sensual delight gasping with desire under the layers of heavy cloth and divine presence. What was a good priest, churchman or abbot to do except banish woman from those places where he indulged his spiritual excess? After all, woman was the enchantress, the beguiler, the succubus of his darkest nights. It was inevitable that the evil of woman, and all that Eve embodied, must be removed from the presence of the one potency that man believed could save him from himself, the hand of God.

The twelfth-century story of Héloise and Abelard is a romance of sensual disillusionment. Héloise's desire reveals the confusion and separation of mystery from earthly love. Her life resonates with the dramas of her era, but also with the soaps of today. Men and women were no different then than they are now, the only change is that we are becoming more aware of the illusion of disparity and more thoughtfully, the parity between us.

Héloise met Abelard when she was seventeen and he was nearly forty. Little is known about her background, but she was a gifted student at the convent in Argenteuil. In twelfth-century France life was shorter, so Abelard might well have seemed a much older man than he would today. Father-figure or not, it seems there was an instant erotic charge between them. Abelard first encountered Héloise as her tutor. He was a learned schoolman of Paris, a humanist, a teacher of theology, and above all a man of the intelligensia of the time. He was aware that Christianity was 'lop-sided', and like the mystic Hildegard of Bingen, sought out the meaning behind the dogma, attempting to include a more feminine image with the divine, or at least a trinity of equality and/or androgyny.

Twelfth-century Europe was comparable to other specific points in history where a sense of change becomes powerfully manifest through culture, literature, theology or intellect. Like the Neoplatonists, the early Greek philosophers, and later the Romantics, Héloise and Abelard were born in a time when ideas, truth and soul were being explored by those who were disenchanted with the cultural constraints of the era. Héloise and Abelard had a sense of the mystery of the soul, and of sensual love. They lived in a era when culture and religion was moving further and further away from the feminine, from the images of the body, nature's beauty and pagan fate. Spiralling out of control, the Christian revolution was abused by individuals with the taste of Jesus's blood on their lips, and desire for power was only quelled through celibacy and piety. But it was celibacy and piety which were the power.

Héloise wrote many letters to Abelard, and it is through these that the story of their relationship runs. There is little doubt that theirs was a relationship of body-soul, not mind-soul, an affair which smouldered for eighteen months, fuelled by passion, inflamed by desire, and which inevitably resulted in the birth of a son, Astralabe. To placate Héloise's family, Abelard agreed to a secret marriage, but Abelard had his own conscience and his own reputation to contend with, and he persuaded Héloise to remain hidden in a nunnery, for her own and possibly his best interests. Héloise's uncle, Fulbert, the Canon of Nôtre Dame, vowed vengeance on Abelard, not only for his lack of celibacy, but also for his abandonment of Héloise, and castrated Abelard. By the age of nineteen, Héloise had taken her vows at Argentieul to please her husband/lover. But for the next eighteen years Héloise wrote Abelard dozens of letters. She grieved at their separation, she battled with the images of carnal passion which invaded her prayers, and even when she was made a prioress, she still believed that her pious reputation was a sham. Abelard propounded her as chaste and the model of piety, and she became popularized as a personification of Mary Magdalene, fallen from grace but risen to virtue.

Abelard admitted he had physical and sexual desire and love for Héloise, and spiritual love for the Virgin Mary, exactly the split that the Christian revolution was adopting. Mystery belonged to God, not to physical lust. A woman was no longer deemed worthy of mystery, and only necessary for the procreation of children. Héloise's sensual world became banished to a convent, castrated as easily as Abelard's genitals. Similarly, modern women can castrate their own sensuality at the expense of careers, intellectualism, feminism, motherhood and spirituality. Dangerous enchantment when suppressed becomes deadly manipulation. We may amputate our own earthly erotic channel to mystery, but we may find that this elegant castration leaves us not only with a wound, but also with a disconnection from ourselves. Héloise reiterates this in one of her letters: 'I should be groaning over the sins I have committed, but I can only sigh for what I have lost'.[3]

Abelard, no doubt unaware that he had a connection to an inner feminine quality in himself, believed truly that his love for Héloise was physically devised. She agreed: 'It was desire, not affection that bound you to me, the flame of lust rather than love.'[4] Desire may draw us together, it is our our initial erotic binding, but unless we can take pleasure in the other senses – our feelings, our needs, our fundamental values – then the relationship may fall apart. The 'erotic' needs to shift, it needs to move into another dimension of sensual awareness to maintain the constant rhythm and interaction between two people. Desire is beholden only to itself, not to those in whom it evokes a response.

For Abelard, lust was sex, and love was something numinous: an approach to relationships we have inherited from the ancient Greek fear of sexuality as power, and love as an ideal. For Héloise her sensual world had been lost to desire and sexuality, but her body called to her, her feelings ached in her, and her soul tormented her as she prayed in true celibate style. Self-abnegating, and denying the pleasures of the profane world she had once so naturally participated in, earthly love was now an illusion. She had been betrayed by it and her senses were only those of torn memories. 'They consider purity of the flesh a virtue, though virtue belongs not to the body but to the soul,' she wrote of the Church. In other words the body can never be pure, it is sensually aligned to the carnality of nature. Yet Héloise believed that taking pleasure in the profane was not a sin, as long as the soul remained virtuous.

Women Mystics

Up until the twelfth century women mystics sought to find the ultimate spiritual experience through union with god. Whether they were merely manifesting a collective need for meaning in life at the time, or were individually searching for answers through the mainstream of Christianity is uncertain. What they were doing was denying their physical sensuality for the sake of grace, compassion and feeling directed solely towards the numinous god-head. Spiritual expectations and the desire for ecstatic union bind us to an illusion, just as in carnal earthly relationships. The problem was that the body became an antithesis to the soul. Soul-mind unified, and body-soul was rejected. Juliana of Norwich, however, a reclusive mystic of the twelfth century, did focus on the feminine in the divine. She believed that the mother was 'the depth of wisdom', and there were 'three ways of seeing Motherhood in God'. Firstly mother as creator, secondly as giving the 'sensual' to our natures, and lastly as seen in the 'activity of motherhood.'[5]

Here Juliana of Norwich and many other mystics like her offer a different sensual grace, an awareness that the sense of compassion pervades an air of mystery similar to that of sexual mystery. Many women who hide their bodies beneath the shrouds of religion are still sensual. Woman's body or not, even though they have denied their sexuality, and chosen a spiritual life through god, or the Virgin Mary, they are still eroticizing their lives through the divine.

Mother Teresa is probably one of the most hidden sensualists of the twentieth century. Her sensual power worked quietly under cover of those grey gowns and religious teachings. It was neither appropriate nor necessary for her to take sexual pleasure, but the pleasure of compassion and the pleasure of a different kind of love, a spiritual union, is as erotic and transforming as any earthly one. It is only when we attempt to rise above our mortality as omnipotent that we lose touch with humanness. We may track the spirit without an awareness of why we have chosen to identify solely with one image, such as god.

A sensually graceful person does not rely on sexuality alone, but a sensually graceful person can be sexual. If however, we over-attach ourselves to transcending the earthliness of being human, it is then we may also trade grace for hubris. Mother Teresa seems on the surface, although she may have underlying motivations, to have kept in touch with her earthly being. Others may have not. The blood of Christ was a symbol of suffering, which many religious women and mystics saw as of great significance. Blood is an ancient symbol of desire locked in our psyche. Desire pulled many men and women into a spiritual life then as it does now. The only difference between desire for sexual love and desire for spiritual love is in the split between the sacred and profane in our culture, not in desire itself. For the medieval mystics and lovers of the Christian faith, safety could be found in sensual grace rather than in earthly love, and for the church, the contradictions in women were quickly contained in the female segregation in nunneries. We danced well with each other then, but it was not to last.

The sacred was now well and truly cut off from the profane. The crucifix had disconnected from the vulva, and was held tightly in the hand, not of god-fearing man, but of woman-fearing clergy. This was the beginning of a cycle in Western mankind's history which was to subject many women to a more terrifying fate.

Fate

Fate is a very loaded word: it implies fortune-tellers and newspaper astrology columns, but its intangibility, like mystery, has long been associated with woman.

Novalis believed that 'fate and soul are the two names for the same principle', and the ancient civilizations believed that woman was symbolic of fate, and subsequently soul. Fate literally means 'that which is spoken'. As 'the Mother of Destiny' fate was known as Mammetun in Babylonia, and later better known as the three Fates, the Moirai – all probably originating from the Indo-European Kali Ma, mother of Karma.

Nowadays we believe we can control the body and ultimately decide our own fate. Fate has become a belief that we can separate ourself from nature, or fate is something we can manipulate, depending on how much money we have on credit. We have conditioned ourselves to believe we can contrive and construct our bodies against the inevitable sense of mortality. We experience fate mostly through strange coincidences, or a sense of destiny when we fall in love. We believe we have free will and then again maybe we don't. Are we to chose our own path through life, forge our way to the top and deliberately pit ourselves as hard as we can against our bodily limitations, or do we vegetate and become couch potatoes? And is this fate or free will? The philosophical implications of fate and free will are too vast for the scope of this book, but is fate's association with women a nagging suggestion of the unknown, the future probability or possibility? For once a man enters a woman's arms, her warm embrace draws him down to her perfume and her kiss, and he is fated by her very presence. Fearful to churchmen, and fearful to all good Christian men, women became the prime symbol of man's darkest fear, the only certain fate, death.

Women as fate is an underlying unconscious tickle in the back of men's throats even now. A man biologically needs to spread his seed, but then finds himself attached and bound by a woman from which he cannot detach himself. She appears to weave her spell, a fatal attraction or an attractive fate, but is never a waste of seed. Desire pulls him inextricably into the web and he is faced with the power of female sexuality to meet his destiny. A woman may keep him in body, keep him from flying into his head or out of his mind into spiritual fancy. She keeps him grounded by her sense of fate. She represents body mortality, and she is all that feminine or yin-sense that he may have been conditioned to reject in himself. He is brought up not to cry, he is brought up not to show feelings, and he is brought up to believe in goals, achievement and potency.[6] So when he becomes fascinated by a woman, enticed and seduced into her arms (not always unwillingly) his own deep well of feeling, his sleeping gods, surge to the surface to play into her sensual domaine. It is unstoppable, uncontrollable, it is awesome and powerful, frightening yet ecstatic.

The Ninja by Eric van Lustbader is a work of violence and stillness, the cross between East and West and the sensual. Here the sensual world manifests as the carnal visceral, violent rawness of nature merging with compassion, feeling and archetypal desire. The hero, Nicholas, is both a man of passion and a man torn by love, or by the world of his senses. But he too like every other man seems to feel that woman is his fate.

'What is it about her, he wondered, that pulls me like a current? Oddly he felt adrift upon the tides. Watching her, the soft rise and fall of her warm body, he knew he was being drawn back to Japan, into the past where he dared not tread.'[7] It seems as though fate, embodied in his lover's body, has taken over his senses.

And what of those men who have chosen a godhead rather than their instinctual world? Those who chose desire for god to prove themselves to be men, rather than their desire for a woman to prove themselves male? Father Ralph de Bricassart, torn between god and woman in *The Thorn Birds,* faces his own hubris and fate in the arms of Meggie.

> 'Because at last he understood that what he had aimed to be was *not* a man. Not a man, never a man; something far greater, something beyond the fate of a mere man. Yet after all his fate was here under his hands, struck quivering and alight with him . . . I am a man, I can never be God; it was a delusion that life in search of godhead. Are we all the same, we priests, yearning to be God? We abjure the one act which irrefutably proves us men.'

Once in the embrace of Meggie, Ralph now knows he could be no more than what he is, simply a man. With 'his cheek against the softness of hers' he 'gave himself over to the maddening, exasperating drive of a man grappling with fate. His mind reeled, slipped, became utterly dark and blindingly bright . . . This was being a man. He could be no more . . . So he clung to her like a drowning man to a spar in a lonely sea, and soon, buoyant, rising again on a tide grown quickly familiar, he succumbed to the inscrutable fate which is man's.'[8]

It is both men and women's 'fate' to find oneself as man or woman through the other. We have to be slaves to each other, we have to be kidnapped and kidnapper, we have to be the bedouins and the wanderers for our senses to burn the mystery through us. There are just as many women as men who disown their feelings and instincts, love or desire. There are just as many women who fear fate as a notion of the inability to control one's life, and to not make one's own choices. Jung said that 'free will is the ability to do gladly that which I must do'. For fate and free will are one and the same. It seems that our deeper intentions and choices are not simply cast from our parental upbringing and our collective and cultural inheritance – our innate fate is something mysterious, both divine and earthly which weaves into us and out of us. Elusively, this energy twists out of reach and beyond us in the moments of complete sensual surrender.

So mystics like Hildegard of Bingen, Juliana of Norwich, St Teresa and many others, both men and women, turned away from the body in search of the spirit. Many turned to Christianity, and away from the sense of the fatedness of being human, to the sense of transcending that human mystery instead. But those who chose instead a pagan spirit or an earthly soul were to become the scapegoats, the temptress daughters of Eve in the eyes of the Church. The sense of mystery was soon to be replaced with the sense of exposure, in the darkest and most potent denigration of women's bodies, feminine feeling and 'incognita's' soul since the elite idealism of earlier Greek writers. The plot thickens, but whose plot was it? Maybe fate's.

NOTES

1. ed. Hayward, p.24
2. Jung, p.103
3. Betty Radice, (transl.) p.42
4. Ibid, p.45
5. Anderson and Zinner, pp.204-11
6. This holds true for many 'Westernised' men, particularly 'the stiff upper lip' indoctrination of Victorian imperialism. Obviously not so much for men in countries where expression of feeling is accepted in either sex.
7. Van Lustbader, p.44
8. McCullough, p.355

PART 4

Sensual Grace or Disgrace?

'Get thee to a nunnery.'

Chapter Ten
'Mary, Mary, quite contrary'

Virgin or Whore?

If a woman is called a whore, is she insulted? If she remains virginal longer than her peers, does this cause embarrassment or delight? But why do these words become insulting or offensive, and can a virgin be sensual or is it a whore's exclusive right? Whore comes from an ancient Indo-European word meaning 'desire'. Houris were sacred priestesses, and the dance of the hours was part of the ritual celebrations of ancient Sumeria. The metamorphosis of the word 'whore' has been largely infilled with the connotation of wantoness and promiscuity by traditional Western patriarchs. Our recollection is more identified with the whores of the past two millenia than those earlier, less contrived notions of women. Virgins similarly were tagged as valued commodity for the purpose of marriage. Virgins were chaste and undefiled, whores were temptresses and dangerous. It was a traditional duty to ensure that the images of women were manipulated for the purpose of progress.

One idealized icon of woman, crafted out of an ancient Mother Goddess image combined with a need for purity, was the Virgin Mary. It may be that this powerful creation of de-sensualized woman by Christian churchmen is the finest paradox of virtuous wantoness. The mother made mother by divine intervention enables her to be a-sensual. Yet at the same time a virgin is man's most provocative seductress, an innocence of the senses waiting to be corrupted by the pious. Mary's imagined smile may have done much to rob passion and sensuality from the symbol of motherhood, but she kept the soul of woman aflame during a period when women were humiliated and denigrated in body, mind and soul.

SENSUAL GRACE OR DISGRACE?

So Who Was the Virgin Mary?

If fate was woman, then so too was the Virgin Mary a religious fate, disguised and dressed by Christian churchmen's own last supper. Mary had been a pagan goddess in many forms before the Christian Church reassembled her into their saviour's Virgin Mother. The ancient mythologies identify her with other goddesses, such as Aphrodite and Isis, born out of the sea, also Nammu, the sea itself, the primordial ocean from which all was created. Her watery nature (Maria comes from the Latin *mare*, the sea, and Mary was always known as Star of the Sea, or *Stella Maris*) was symbolic of the feminine feeling nature. Her earthly connection, as '*Mater Virgo*, or virgin matter', the 'unploughed soil', was the base material of our substance of being, unformed, and like a new shoot, ready to grow into life.[1]

Virgin Woman

In Roman times, the vestal virgins were priestesses who needed no man and never married. This does not mean they were celibate or chaste. There is strong evidence that these temple priestesses, like the 'houris' and the sacred harlots of Babylon, provided sensual pleasure for those who would chose the sacred path through sexual practices of the divine.

But early Christian leaders redefined a virgin as an exclusively chaste woman, untouched, uncontaminated by her own sexuality and men's. However, the intrinsic value of her nature, which the Christian Church chose to overlook, included her self-containment, her need of no man and the suggestibility of her latent fecundity. The early Church also excluded the vestal virgins' earlier connection with sacred pleasures. Mary was better a virgin than a mother, for mother implied carnal knowledge. A virgin now only implied chastity. Sexuality was explicitly found in the woman who identified as a whore. Or was it?

Fate and woman, as embodied in Mary, still wove silken threads around mankind. Her ancient sensual and sexual connections were closer to her heart than was realized. The Church may have created an idealization of the feminine, a woman who defused their own insecurity with fate, but Mary still constellated the power of the Moirai as she sat weaving her thread. She embodied the triple goddesses themselves: she was all three Marys who stood at the foot of Jesus's cross.[2] Mary's ancient past was glossed over and reworked into the Church's songbook. She had become their one feminine image that would ensure the continued control and containment of women, for no female would ever be able to live up to this ideal. Mary 'alone of all women is blessed, because she is virgin and fruitful, she conceives in holiness and gives birth without pain.'[3] A few centuries later, Bernard of Clairvaux would say the Virgin Mary is 'warming our hearts far more than our bodies, fostering virtue and cauterizing vice.'

The Split

Up until the twelfth century the fervour of Christianity was embraced equally by both men and women. Sex was considered sinful, and guilt flared through the nostrils of churchmen and nuns alike, until the eleventh or twelfth centuries when priests, theologians and mystics became celibate to avoid the terrifying Boschian reminder of hell and eternal damnation. The inherent belief that had grown out of the Greco-Roman confusion about men and women was instilled heavily in the mind. Pandora and Eve, and therefore all women, were potentially sinful, sexuality their worse expression. Misogyny became a black boil waiting to be burst on the neck of the Church, particularly in Europe. Earlier pagan spiritual worship dived underground, and further afield, Middle-Eastern Christian doctrines like Gnosticism took a different attitude towards the image of the feminine.

For Europe, the lust a woman inspired in men, particularly one might imagine churchmen, feverishly panting on their pulpits, was now becoming fashioned as the ultimate root of all evil. Earlier leaders, such as Clement of Alexandria in the fifth century, suggested women wear veils and men grow beards to emphasize their sexual differences. Many Middle-Eastern countries still adopt this style to express their religious belief, but woman, hiding beneath her black veils and gowns, still emanates a sexual power. The one reminder of pagan sexuality, disguised with the trappings of male-dominant cultures, can never be totally suppressed and often manifests itself as the power behind the patriarchal leaders themselves.

The philosophers, thinkers and writers of classical Greece and Rome had left their own philosophies and writings clearly defining men's attitude to women, one which slowly grew into the collective European mind. Ordinary men and women still desired, lusted, separated, hated and were rejected. Ordinary men and women were sexual, sensual, carnal, but if they were to look towards the Church for their spiritual unfoldment and inner sanctuary, they would have to conform to the belief that sexuality and the pleasure of the senses were incompatible. You could marry, so that the sin had purpose, or you became celibate. The good wife was merely essential for the procreation of children, and the denial of sexuality, rather than its pleasure, was a road to spiritual union with God. Christianity's and individual man's own weakness leapt across the abyss towards potency. The vulnerability and emotional heel in every man chose to exploit the power of God.

One of the most influential writers in this repression of both woman and sexuality was St Augustine. 'I know nothing which brings the manly mind down from the heights more than a woman's caresses and that joining of bodies without which one cannot have a wife.'[4]

Rebirth

But among ordinary people, the noblewomen, the dutiful wives, and the mystics and Béguins of the Middle Ages, earlier cults around Mary as the earthly mother of Jesus had taken root. Out of their ancient, forgotten, pagan past, the popular imagination had brought back the Mother Creator. During the first few centuries Mary was more a mother figure of Jesus than a spiritual exercise. To be simply the sensual, earthy, compassionate mother of a male saviour was a deeply compelling image. The earlier goddesses of ancient Greece and Rome were still alive in Mary's breasts, not exactly where the Christian church had anticipated. How could a mortal woman, menstruating and sexually bewitching, be the place from which Jesus came? How could they defuse the Mary cults, and keep god unlusting and intact?

But the Christian god needed an ally, a feminine image that upheld the view that woman's sexuality was damned and damning. Bring in the Virgin Mother archetype, pure and undefiled by sex. It was fundamental to Christian dogma that Mary was pure. No longer just Mary, no longer a woman capable of sexual relations, sacred pleasure was now seen as the undoing both of men and the Church's fount of power. Spirituality flew off the ground into the abstract with air-borne angels, far removed from the ancient divine found through nature. A woman must be created to transcend all women. Mary was born again. but this time cast in a solar fire,rather than a moonlit night, stripped of the lush fertile chamber of her earlier sacred sensual world. 'In the very celebration of the perfect human woman, both humanity and women were subtly denigrated'.[5]

Desensualized Mary

As both virgin and mother, immaculately conceived without sin, Mary was humble, chaste and pious. Marian legends demanded that Mary was beautiful, but no man could ever look on her with desire.[6] Take away desire and you have the compassion and gentleness of woman without the wildness of the feminine aspect. You may no longer fear woman. The imagination of artists, writers and mystics attempted to dehumanize and mostly to desensualize her. Her body was covered, her halo distorted into the most ephemeral of shapes and illuminations, and her face painted as merciful, unseductive, or benignly pious. Had they finally created the image of a woman who could never launch a thousand ships? Could this perfect woman be neither temptress nor beauty? Was this a woman who had no hint of sensual suggestion, or did she still suggest the nature of Eros to those who knew the intimacy of love? How can you paint a portrait of woman suckling a child at her breast without it being perceived as sensual? A fifteenth-century Dutch painting is one of those rarer depictions of Mary feeding a child at her breast, presumably Jesus. She gazes as if enthralled by the suckling. The erotic contact between child and mother is established, delicious and womanly, the only contribution to her spiritual unworldliness is the requisite halo. Whoever modelled this Madonna modelled a woman, flesh, blood and sex, not a misty image, a delusion of contemptuous clergy. The only elimination of the sensual in Mary was by the Church chauvinists. The artist saw the woman first, and the icon she represented to most people was one of the ancient symbols of the feminine. To the people, Mary was sensually complete.

The Breast

We still have this problem with breast-feeding in public in Western cultures. Men and women often turn their heads in embarrassment or discomfort, their reactions possibly indicative of their own early breast-feeding routines. Don't we all respond to that primal basic image of our own sensual needs when as newborns we are the masseurs of our mother's breast, the kissers of her nipples, the ones who cause her body to be at its most voluptuous and erotic? Breasts are both life-givers and sexual arousers, but it is the mind which plagues the nipple with guilt. We are charged with an erotic connection when we indulge in the nipple zone, whether men or women. Men find women's nipples comforting or arousing, women find their own nipples comforters, arousers, and self-arousing. We make a third out of the two, but then women are triadic it seems. Mary, Mary and Mary played Scissors, Paper, Stone around the base of Jesus's cross; the three fates spin our weave, and the triad of Virgin, Mother and Crone is an ancient symbol for the yonic triangle, or vulva.

And what of men's nipples? Some enjoy the pleasure of their own, others can't bear to be touched there, nor can they feed a child. Ingrained in our personal psyche is the memory of breast as comfort and life, but ingrained through social inheritance is the forgotten but yet suggestive memory that woman's breasts are also our vice. Vice is derived from an ancient word meaning 'into two', or 'asunder'. Two breasts, and the symbolic equivocation of the feminine principle neatly displayed across every woman's chest. Ironically, the Virgin Mary embodies the duality of woman as both virgin and mother, another link in our sexual split.

Cult of Cults

The cult of the Virgin Mary is the cult of cults. However religious we are or aren't, the images instilled in our cultural legacy are like implants of an ideal woman that women are incapable of becoming. The more we are mother, the more merciful we are supposed to be; the more we are virgin, the less sexual we are, and the only role where the sensual satisfies humanity is in the lover or the whore. To placate the popular view, and to ensure that women of the church and mystics alike were vowed to chastity, Mary became fashioned as the ideal. 'The Immaculate Conception remains the dogma by which the Virgin Mary is set apart from the human race because she is not stained by the Fall . . . As the icon of the ideal, the Virgin affirms the inferiority of the human lot.'[7] The Virgin Mary was a profound concept which duped the world for hundreds of years, and simultaneously let down the coils of women's hair across the churchmen's pillows. The sensual world was as dramatically alienated from the sacred world as it had once been part of it.

A woman who was a mother was now considered sexual and and in need of salvation. If she was 'with child' then she was also 'with body, fate and sex', her sensuality alarming compared to the pure haloed images of the Virgin Mary. Even women mystics described Mary as 'resplendent' and agreed that she enhanced the image of woman, rather than debasing their sex. Saint Bridget was the first to agree that knowing Jesus was 'where there is no carnal pleasure'.[8] It was not that the Virgin Mary did not promote the goodness, compassion and preservation of femaleness, but she also took away the very fruit, richness and desires of woman's mystery. She became an abstract display mannequin, a man-made woman who had been crafted skilfully from the broken pieces of fallen goddesses and their dying culture. The decadence of woman's fall was still coloured with lunar light, but the shining haloes of the Virgin Mary were the symbols of the Church's own solar principle.

Aim, project, and fire your ideology into the very womb of the people, where it is sure to take seed. The irony was that the very body and soul of woman which the Church debased was the very body and soul into which they delivered their dogma. By imbuing Mary with virtue (meaning literally manhood, or masculinity) they had not only pumped masculine doctrine into the womb of the Western world, but also their own insecurity, their own terrifying vulnerability in the arms of a woman. Both men and god still depended on mother, or Mary, to get them out of trouble. A mother-figure was needed then as it always would be, but she was tied to a fantasy stake by the many who would eventually exploit the power of Christian philosophy. Searching their way through the darkness, this was a millenia whose religious collective relied on hope and faith,[9] rather than belief and being. An antidote was needed.

Mary Magdalene, Lover of Jesus?

A pattern of sensual repression was being established, but it needed one more finial, one more flourish on the Doric column of Christian faith – the twist of the redeemed whore. Virgin and whore became the two most split images in the Church's doctrine, and so it seems in women too. If mother was being lost to the virgin, then woman too was losing her sense of identity.

Every child has a mother, but the child of God, Jesus, was not only uncontaminated by his mother, but also untouched by sexually active women. Biblical stories abounded about Mary Magdalene, her compassion, her suffering and her love. The best of the myths are from Gnostic scriptures, which tell of Mary Magdalene as lover of Jesus and mother of his child. Biblical stories tell of her as a repentant whore, but it may be that the profundity of these very different versions is in the actual controversy that has always accompanied her image. It is uncertain what her true role was, but as a symbol of fallen woman she was remarkably overt. Magdalene means 'she of the temple-tower', and like the Virgin Mary, her ancient associations seem to reconcile her as one of the Triple Goddesses. The Gnostics believed the 'Marys' were one primal feminine power:

'I am the first and the last.
I am the honoured one and the scorned one.
I am the whore, and the holy one.'[10]

Mary Magdalene may have been another emanation of 'the first and the last'. Like the Virgin Mary, a necessary otherness; a version necessary to complement and emphasize the chaste one, suggestively sensual by her own rather ambiguous legends and stories. The imagination of mankind enchanted her into myths and tales, laced her with intrigue and mystery, the sure sign of temptress. Ironically, this illusion of the sensual as intrinsically wicked is the one that became indigenous to Western culture. Cast Mary as serpent and Eve, cast her as mysterious and yet earthy, fecund yet erotic, and you have the archetype redeemed in different guise.

In one sense Mary Magdalene was the new, earthly dark feminine cast into the heavenly pure Christian world. For the purposes of the Church Mary Magdalene was redeemable, the whole doctrine behind the Saviour. As a symbol of humanity, her depravity, her wantoness, her very femaleness represented the sins of the world and mankind, and so it was that Mary Magdalene fitted into the plot quite neatly. If Mary Magdalene was the mistress of the resurrection, then she was also the mistress of Lucifer, and was known in medieval myth as the 'light-bearer'. She was an icon of grace and feeling, an imagined embodiment of humanity's painful wound around pleasure, now redeemable.

In the Bible she was repentant whore, but to the Gnostics she was Christ's beloved companion. One Gnostic story suggests Mary and Jesus were lovers, the passion so strong between them that they hid hermit-like in the mountains to consummate their love. Here their sexual relationship revealed to the disciples what the world truly needed – love. 'But Christ loved her more than all the disciples and used to kiss her often on the mouth . . . They said to him, "Why do you love her more than all of us?" The Saviour answered and said to them, "Why do I not love you as I love her?"'[11]

The New Sense

The end of the Dark Ages saw the beginning of a new sense. The shifting roles of men and women as they danced on the divine loom saw to it for the next thousand years that the lustre of the senses was to become an antithesis to the tarnish of the powers that be.

Cultural assumptions about gender equality still persist through our language, and we still make judgements about ourselves and others without ever thinking about the true meaning of the senses. The same illusions still persist. If a woman is sensual she is usually considered whore, lover, *femme fatale*, disgraceful, seductive, alluring, beautifully dangerous or mysterious. If a woman is sensually aware, she is considered unlikely to be a mother, and certainly won't be a virgin. If a man is sensual he is considered a freak, effeminate, artistic, mummy's boy, weak or over-sensitive. If a man is sensually aware he is usually considered to be gay, a woman's man, a lover or a 'romantic', and he certainly won't be a man's man. But there were those who doubted the indoctrination of this new sense of apparent differences. The sensual had to move on, move away and burrow deeply to survive the next few hundred years of female subordination in Europe. So where did it go?

The sensual still lived deep in the hearts of many ordinary people, it shifted into literature, art, paganism, music and legend, and then with a profound backlash that seemed to rise woman into the status of lover and beloved, it scrambled itself into and out of the world of courtly love. This rich, erotic and intensely powerful infusion of love into Western culture quietly shattered the collective nerve, and love, earthly passion and the imagination were once again for a while the pathway to the divine. But the wires were to get crossed.

> *'Take your aloes and your perfumes, and leave my sight. My eyes see only the fallowed fields and the fruits, not the light and the likeness of God. Be gone soon or be sacred; be pure or be purged; be virgin still or be viced. You came to my bed an innocent, you leave it a woman, but you alone are damned for you brought with you the serpent of my desire. Yet for now Rest sweetly in my bed. You who have bedazzled me with the jewels of Jesus's mother.'*

Virgin, Whore or Both?

Recklessly, the Church and their cult of the Virgin Mary were closer to honouring and worshipping sensual and sexual woman then they ever admitted, either to themselves or to the world. The archetypal whore is provocatively sensual, she knows the language of sexuality, she knows what it is to be embraced by a man. Women are all whores once they are no longer virgins.

But a virgin is *suggestively* sensual, she is innocent, naive, chaste, untouched by men. She carries the mystery of woman, not the experience of sexuality, only the awareness of it. She may be the greater temptress, able to destroy a man's soul while she remains pure in body. Innocence is corruptible, and the sense of chastity is highly arousing to a man's sense of potency. Virgin is sense delight, whore is sexual compulsion. To the Christian Church the Virgin Mary was an attempt at defusing the 'sin' of the sexual woman into a paradigm of virtue. Yet the paradox was that the Virgin Mary carried not only the experience of her motherly past loaded with the baggage of ancient fertility connections; but now she was imbued with virginity, an alluring erotic energy. The Virgin became an icon for many men to rub themselves against both sexually, and emotionally. She could take a man's soul without him having to spill his seed, and the seductive image of the Virgin shimmered before men's eyes for the next thousand years like the girdle of Aphrodite had once done before. Sensuality was exclusive to women because the Virgin Mary was the pure icon of woman-ness, an easy reflector for men's own repressed sensual needs.

Salome

The legends surrounding Salome merged into her personification as a carnal virgin temptress by the end of the nineteenth century. The brief tale in the Bible trivialized some important aspects of Salome, seeing her only as the dancer who pleased Herod. She has been identified as Mary Magdalene, as present at the birth of Jesus and at the death of Jesus, not forgetting her involvement in the death of John the Baptist.[12]

As the dancing daughter of Herodias, however, she portrays the paradoxical image of the enticing and dangerous, yet innocent virgin. The decadent artists and writers of *fin de siècle* nineteenth-century Europe became obsessed by the duplicitous lure and chasteness of virgins. Moreau's painting 'Salome' is indiscreetly 'sensual'. Her pose is teasing yet coy, her dance enveloped by sparkling jewels, silk and gauze, her body wrapped for pleasure yet rapt with the innocence of a virgin's tease. Her gloating face demands the head of John the Baptist for her scheming mother, Herodias. But was she truly innocent or lusting for blood too? Flaubert in his tale *Herodias* takes Salome into the ethereal realm of chaste wickedness as she dances. She is the virgin whore herself, as powerful an image as the Virgin Mary, cradled by purity, yet her female presence sexually undoubted. Salome dances suggestively of the sensual embrace she has yet to experience, she is the dual nature of the longed for image, both corrupt, yet incorruptible. 'Without bending her knees she opened her legs and leant over so low that her chin touched the floor. And the nomads inured to abstinence, the Roman soldiers skilled in debauchery . . . the old priests soured by controversy all sat there with their nostrils distended, quivering with desire.'[13]

It is precisely the image of the virgin who is so self-possessed, so unblemished by her own sexuality, and in need of no man, that many men fear and desire. To deflower her is to have power over her. To take away her self-possession, to demand of a virgin her chasteness, is to remove her autonomy, her sense of beingness. The Virgin or the Whore, and mostly the Virgin-whore, is the one archetypal image of women which seems to persist in the collective psyche. The profound dichotomy of life is symbolized by this split in our imagination. The interplay of yin and yang is not the separation of opposites, but the dynamic and tense relationship between both, the perception of which is simultaneously the same. If we can perceive the qualities of virgin and whore as a dynamic whole, then we may begin to understand how this split is embodied in woman.

And What About Zeus?

So women are Salome and Eve, Mary Magdalene and Pandora sealed in the jar of life. Locked in the symbol of the Virgin Mary's body, and her divine conception via god, is the secret relic of the sensual interface between the profane and the sacred. Zeus, like the Christian god and many other global and mythical gods, fathered semi-divine beings. Zeus omni-entered the nymph Semele when she demanded of him divine passion. But Semele's reward was to be ripped apart by his potency, the flames of his sacred desire devouring her in an inferno. From this union Dionysus was born, the god of carnal pleasure, his embryo saved from Semele's burning body, and sewed into Zeus's thigh like a flap of flesh, a wound of remembrance. Dionysus is he of the instinctual chthonic world sense. He is the god who is 'twice-born', saved as a baby by Rhea from the revenge of Hera, and reassembled into life. Dionysus takes us to the moment of choice, he is the god who teaches mortals the transience of pleasure and pain and of the ecstatic moment of sexual release.

Similarly, with the Virgin Mary's union with god a new symbol is born. Dionysus embodies disorder, confusion, catharsis and instinctive pleasure. But this Christian embryo became an easy scapegoat for the feeding frenzy of lust denied, hungry Church despots who needed more than a virgin mother, and a Saviour son – it needed real women on which to feed. Dionysian cult worship and rituals provided a containment for cathartic sexual and earthly experience, but Jesus as an icon fares better as the son of simple folk, as one who loved Mary Magdalene, and whose midwife was Salome. To be awarded an immaculate virgin mother and a divine father meant you were destined to be above contamination, a symbol of guaranteed atonement. This time not sewn in a God's thigh, not hustled away as a child to become a symbol of chaos, but as a sacrificial symbol of assurance. Civilization, embracing the Church, followed the golden solar principle and the redemption of mankind. Dionysus was now truly abandoned for the golden boy, so the culture of sexual love, pleasure and the real interface of earth and sky dived underground, back into the pagan world from which it had emerged.

NOTES

1. Watts, p. 107
2. Ashe, p.135
3. de Voragine, p.205
4. St Augustine, cit. Mann and Lyle, p.34
5. Warner, p.xxi
6. De Voragine, p.152
7. Warner, p.254
8. Zinner and Anderson, p.219
9. Faith is subtly different from belief, in that the former implies putting one's trust or belief into the statement of another, whereas the latter implies an inner intuitive knowing which does not rely on the words of others.
10. Nag Hammadi Library, p.137
11. Nag Hammadi Library, p.138
12. Walker, p.886
13. Flaubert, p.79

Chapter Eleven
Feeding Frenzy

'If ye perceive a sudden sweet taste in your mouths or feel any warmth in your breasts, like fire, or any form of pleasure in any part of your body . . . then this sensation is very much suspected of coming from the Enemy; and therefore were it ever so wonderful and striking, still renounce it and do not consent to accept it.'[1]

Sensual Suppression

To deny ourselves the expression of sensuality means we may deny our body its connection to soul and nature. Even before the Dark Ages we had been fed with the belief that a woman was not only evil, but against God, and that sensual pleasure was a manifestation of 'the Enemy'. The expression of the sensual was not denied, it was suppressed.

Sprenger and Kramer's odious yet powerful and cunningly constructed misogyny, the *Malleus Maleficarum*, published in 1486, taught that women were possessed by the Devil, women fornicated with Satan, and 'woman is more carnal than man'. She is 'insatiable' in her sexual desires. Church leaders, in their distortion of Christian belief, had run out of rich heretics to bring to the Inquisition, so they turned to the common people and country pagan worshippers. Their own repressed pleasure-dome groped and snarled in the darkness of Hades until it resurfaced as demonic, the very Devil they believed possessed woman. If women were known to make love potions then they would inevitably make impotence potions; if they were there to heal and cure, they were there to destroy, and specifically to destroy God. If woman was to take pleasure, she was there to give pain.

Europe became a splendidly grotesque Kingdom of God, a torture rack of sexual inhibition and neurosis. Local villagers accused each other of heresy, of witchcraft. A soap opera of diva proportions was scripted around every village pond as women were drowned, burnt or hung. The perversion of group fear that pervades in small communities became a discharge of odious soma, not of the Gods, but of a guilt-ridden ejaculation in a mirror. There was a mass air of guilt and of accusation. If you were not a witch then you found one to accuse, if you were accused you began to believe you were an associate of the devil. You said things in the torture chamber because you had no choice.

We do this today in our relationships: we accuse our partners, and accuse ourselves. We blame everyone else for our dilemmas, demand compensation for mistakes valid or not, both sexual and material. Many men blame women for their fate, and women now blame men for theirs. The richness of the inexpressible nature of the body-soul has been desensitized in a similar feeding frenzy. Sensual and sexual pleasure has become a bondage scene in a porno film. The torture chamber eerily communicates a perverse sexual turn-on, just as it probably did five hundred years ago, the only difference being now there are lights, cameras and the chance to view such delights in your own home. We each have our own scaffold for the goriest of sights in our living-room, whom we place upon it depends on our current Godhead. Five hundred years ago it was imagination and fear that disposed of pleasure from sexuality's arbour, not explicit sado-masochism.

Up to the seventeenth century, 100,000 or more witches were recorded as being strangled, drowned, burnt or beheaded, and there were probably thousands more. The evil of woman and her inferiority was now fouled by the Devil disguised as sexuality. But the demons that the Church believed controlled women were the demons of their own darkness, not of woman's body. Suppressing the power of woman's sexuality meant the suppression of men's own desire-nature, which was not compatible with the godhead the Church had created. The phallus must be a power in its own right, it must not be handled by woman. Instead, a woman was to be de-witched of her wisdom and sexuality, with the fire, the water or the Churchmen's halitosis.

Midwives were particularly prone to visits from the Inquisition. Birth was still mysteriously a woman's power, and those who assisted in the delivery were often accused of snatching children and babies for devil worship. Religious leaders opposed all ancient beliefs, and if the pagan world was threatening, then woman was the prime symbol of it. She could be lured more easily by the Devil, for like Eve, she was 'feebler, both in mind and body'.

This Christian anti-sensuality hunt spread further afield than just Europe, for wherever Christianity went with its missionaries and reformers, so too did the doctrine of displeasure in the senses, rather than its pleasure.

In cultures where the female principle was honoured equally and remained untouched by the European neurosis, there was less trouble with sexuality, the body was revered and sex was the celebration of sacredness. Egyptian women chose their husbands and lovers, while sensuality was a joy in Hindu religion as women embodied the power of the goddess Shakti herself. Through union with a woman, body and soul were united in Taoist philosophy. Similarly, in Far-Eastern cultures women and men were not denied their pleasure, rather they were encouraged to indulge in the sensual world.

Sensual Heresy

Churchmen and religious leaders turned to hunt woman. From Paul and St Augustine's teachings evolved a dangerous attack on the evils of the female. 'It is good for a man not to touch a woman. Nevertheless, to avoid fornication, let every man have his own wife. But if they have not continency, let them marry, for it is better to marry than to burn.'[2]

From the mouth of Thomas Aquinas, and many more, came a rhetoric of belligerence and accusation against women. 'As regards the individual nature, woman is defective and misbegotten, for the active force in the male seed tends to the production of a perfect likeness in the masculine sex; while the production of woman comes from defect in the active force . . . or even some external influence; such as that of a south wind.'[3]

But any cultural suppression of bodily pleasure seems to result in perverted ideas, cruelty, lethal manipulation and misrepresentation of its true nature. Our own expression of our sensual nature becomes distorted. In many men the gradual belief was that it was safer to deny sensual pleasure altogether, and impotent, deadly and dangerous to remain in a woman's arms any longer than necessary to plant his seed. Women were indoctrinated into the belief that their sexual power was evil, and that it threatened to undermine the Kingdom of God on earth. The Church's own desperate neurosis with their sexuality was displayed through the torture chamber. Whatever went on in those torture rooms we do not clearly know. There was contradictory evidence from the tortured and the torturers. But were those who tortured any more rational than those being tortured? The rational versus the irrational is a cultural fixation which continues today.

Sensual Warrior?

An early 'agent of the Devil' was Joan of Arc. Pious and virginal, she was also a great warrior, a restyled Amazon or Artemis, devoted to the Virgin Mary. Joan was impressive from a male perspective: she was fearless in battle, and knights who escorted her are recorded as saying they felt no sexual desire for her simply because she displayed no sensual messages. She became discredited by the English who believed she had 'enchanted' and beguiled them on the battlefield. Her identity was anchored around the masculine image, she wore men's clothes, and her suppression of her femininity continued in prison when she refused to wear women's clothes and chose the glorious martyrdom of virginity over a life-sentence which might have included unwilling sexual encounter. Without her virginity, she believed she would be abandoned by God.

But did Joan of Arc embody the womanly sensualist of our collective image, or was her dedication and commitment to her religion a different 'sense' of compassion? The virgin warrior is an image that we recognize even today. Feminism and the need to keep up with changing values have led many to chose careers, financial success and girl power over their instinctual or feminine nature. This does not mean that one cancels out the other, but it seems that the choice to identify with freedom and equality means women often deny or demote the realm of the body-soul for the realm of the mind. Women have also begun to be embarrassed about the mysterious sensual quality they are perceived to express, while men still feel smugly threatened by it, whatever IT is.

Joan of Arc's compassion for all that the Virgin Mary symbolized, in whatever flavour it returns to us now, is the tenderness of the feeling sense. But alone, this sense of feeling struggles. Without manifesting feeling through relationships, compassion becomes a heavy burden. Taking pleasure in the body means we celebrate and eroticize nature and the world soul, both through the material and that which is beyond it, the unknowable. Body and soul are inseparable, and it may be that if we suppress bodily pleasure, the sacredness of nature's energy cannot surface. Caught between spiritual belief and masculine honour, Joan of Arc personifies the split in the collective values of those ages. For not only was there a distinct gap looming between the images of Virgin and Mother, but men's choice of God's material power or God's spiritual one, was a widening crevasse. Joan of Arc chose only virgin warrior, at a time when woman's sexual power was turned retrograde under the light of a wolf-moon.

The Inquisition was a fearful hatred of self seen in otherness, but it became a grand-scale inhibition of women and all that they symbolized, especially the heretical expression of sensuality. Heresy originates from a Greek word meaning 'to chose for oneself'. If the women and men who believed in pagan religious practices and worship could not choose for themselves, then sexual passion, desire and love could not be chosen for oneself either. Sexuality was heresy in the eyes of the Christian patriarchs, and so was its pleasure. But it was still not too late for earthly love.

And What of Earthly Love?

Profane pleasure had already been creeping seductively through the courts of medieval Europe in the guise of courtly and romantic love. God could not redeem man for coming from woman's body, but he could redeem him for his avoidance of her. Heaven was seen as the antithesis of sensuality, and hell a more suitable place for lasciviousness and the witch scapegoats. But the sensual was to be saved. Those who found beauty in art, poetry and music were those who had charisma, nerve, feeling and who took pleasure. They were men and women who sensed more than civil hatred and power, they sensed Peitho's enticement and Dionysus's panthers. These were heroes and knights, warriors and spiritual guardians, singers and poets, noble ladies and courtiers, awed by something beyond potency and power. They knew of Grace, and of Love, and that She must take them to the Towers of Babylon to learn the sensual art.

Aphrodite listened to the reconstructed voices of all the gods and goddesses in one god. She evoked mobility, she charmed and wove her own spell with another more potent love, that of Eros's passion. It was this love potion, just like Tristan and Iseult's that turned men again and again to woman; to find themselves in her, to find the divine in her, and to celebrate secretly their embrace – while the witches burned. The cult of romantic love lay side by side with the cult of God's punishment, 'drown and be innocent; float and be guilty.'

20th Century 'Sink or Swim'

Even now, at the end of the twentieth century, it seems we still unconsciously carry the thread of a witch's hair and the sink or swim dilemma. Maybe there are some of us whose ancestors were accused of witchcraft or heresy, and those of us whose ancestors were inquisitors. Innocence and guilt about sexuality are still rammed down our throats, sometimes rightly so, sometimes not so wisely. But taking sensual pleasure seems to be a forgotten aspect of our sexual world, or it has become devalued into the sleazy, the porn and the pickings of media crows. Innocently we drown in the media dogma or guiltily swim towards another god with a daily voice: become rich, become a millionaire, we must breed and there a million new ways of doing so. Guiltily, we take pleasure alone, or read manuals on how to be sensual, forego human feeling and indulge in perversions or orgies. We follow the Maenads' dance with Dionysus, ecstasy and sexual gratification, the sensuality of hedonism taken to an extreme, but with no gain, for we always believe we must gain. We work out at the gym to look beautiful, then use bizarre sex toys and fall into complicated or ill-met-by-moonlight relationships in an attempt to satisfy our sexual longings. We are civilizing our natures by day, imbuing them with Bacchanalian excessiveness by night. We have adapted the inhibition and heresy of the medieval sexual impulse into a new version. Heirs to any ancestral hall always carry their family's curse as well as their purse.

There are those who are inhibited, and those who are not. An inhibited woman may dial every help-line, an inhibited man becomes a flasher or a child molester. An heretical woman has diverse relationships, men and women, and is usually faithful to all; heretical man flounders in supposed phallic omnipotence, but equally feels too deeply when touched. It is a sting, the eccentric orbit of his perception of a woman's touch which reminds him of his witch-hunts long ago.

The Uninhibited

If there were to be modern-day stereotypes of the uninhibited sensualist and the inhibited one, they would probably resemble the following examples. Uninhibited woman meets her lover and dances naked before the log fire until dawn. She'll seduce, curve, drain and kiss her lover to pieces, and then when she has caught him she will train him, ensuring that all his own instinctive attempts at copulation have been civilized out of him with her uncivilized seduction. Then she may become temporarily inhibited, reclusive, look for work, escape or find another lover. She is unpredictably wild about every new man she meets in the corridor – is he inquisitor or troubadour? Can she tempt and enchant another, or does her own autoerotic nature move her enough to live as a hermit and do without him? Uninhibited woman finds the monotony of intimate familiarity a chore after a while, but equally she needs to relate to experience the sensual.

Inhibited woman, on the other hand, has a night out with the girls to watch male strippers. Inhibited woman usually stays at home and loves her husband. She goes to the beauty parlour, prefers her sexual pleasure to be under the cover of darkness, her own desire fluctuating between rampant and withdrawal to the far side of the bed. She blames her hormones, he blames her for being a woman and shuts himself off from his sexual urgency in one form of addiction or other. If he evolves into inhibited man he may then sneak into strip clubs, porn shops, or find a prostitute. If he reverts to uninhibited man he will take a lover, stop at nothing to feel sensually loved, sexually desired and then begin to fear the dilemma of which woman, lover or wife? How can one woman ever encompass both?

Mother or Lover?

Mother woos us first. She is both Aphrodite and the moon, our erotic grounding to earth. She must seduce us into love, wean us away from whence we came and our barbaric savage babyhood. She is both our darkness and our light. For a woman to become a mother means she slips out of her dangerous dancing shoes and into maternal lushness. A woman knows her body is comfort and instant pleasure to the child she conceives. But where has her vixen-eyed, exotic, wild-woman-lover sense gone? Collective expectations and perceptions have categorized and compartmentalized women as either mothers, lovers, virgins or whores. Because we are definers, measurers and observers, and wish to split more than the atom, we have created the image of women in the genre of one or the other. But a woman is exactly all these images. Woman is virgin, lover, whore, and how could she be mother if she wasn't the others? Our insecurity about the tangible ambivalence of nature becomes safer in the knowledge that lover becomes wife, and wife automatically becomes mother. But what about the mother who is a lover, and the whore who is a virgin?

Woman's Dichotomy

Some women are fearful of sensual pleasure and deny and repress the joy of body love. Becoming a mother to end one's 'sexy' days is tempting to many, but the social conditioning of how we see the image of 'Mother' means women often feel guilty if they still carry a sensual image. When women make excuses about their figures sagging after pregnancy, they are usually responding to social stigmas about weight, fatness and beauty. Those that are thin-waisted and tanned are apparently fashionable, those that are overweight, pasty, and dowdily dressed are not. True inner beauty is the soul of woman and man, whatever shape, colour, width or weight. It is our relationship to our sense of beauty, coupled with collective assumptions about what is in vogue or not, that disturbs the eye of the beholder.

Take the case of a woman, twenty-something, sleek, slender, successful, and enjoying the enchantment she exudes while being the creatrix at work, or Aphrodite down at the wine bar. She dresses and adorns herself, sings and smiles as she perfumes and oils her skin and her words. She is not a virgin, but she is a self-possessed autonomous being and in no need of a man to be her breadwinner. Yet she flirts, teases, seduces and throws her head back the first night she meets him, her lovely white neck revealed as she twirls her long black hair above her head. The sexual symbolism raises his eyebrows and his blood pressure. Some archaic image of the taste of blood, the power of possession and vampires overwhelms him. She is enchanted with her own enchantment; he is lured by her sensual game as she licks the bottom of the glass, her tongue curving around the stem as if she must have him.

Some years later, she has given up courting pleasure, the thrill of romantic dates, of love, desire and languid pleasure. She has surrounded herself with tiny children, another sensual world of nappies, vomit, pregnant indulgence and sleepless nights. Her relationship to him is different. She is tired and mostly doesn't want sex. He is frustrated and unconsciously jealous of the way she seduces her babies, not him. She has become immersed in plastic toys and rubber gloves, her body no longer a mystery to him as she struggles with lochias, sore nipples or mouth ulcers. And then there's that vision of his mother which keeps slipping into his head as she slips into his bed. He was right, woman is the earthly darkness he imagined, only now she is also becoming his mother!

So she is now really mother and sometimes lover, and when she has time to recall her own desire, her own need for fundamental earthly touch and tenderness, she may play the diva of love. Her own dilemma is that she is expected by him to be either mother or lover, to be both is incongruous. To him, she has not only transformed into a real-life brooding hen, but she has eerily begun to epitomize mum.

Mother's Wooing

Wooing is to court, to exchange, to lead into and out of oneself through another. Men need to be wooed by their mothers, as much as women need to be wooed by their fathers. This is the language of love. We may not speak the same language of love as our parents, and this is where the problems in later life often repeat the patterns of these first triggers. It is not our parents who are to blame, for they are profoundly enmeshed in the interchange of our perception of the world and are often simply early catalysts for our own burnt cakes.

If we have received or exchanged the wrong language about our sensual needs early in life then we may as adults deny we have a place for sensuality, which we find difficult to integrate into our relationships. Overt repression of our earthly, dark, squalid side which loves goo and muck may meld with the parental voice that says, 'Don't be dirty, don't be disgusting'. Our own personal voice would rather hear, 'Enjoy it!' There are parents who are offended by filth, dirt, bugs and demanding huggy children. They prefer a brief kiss on the forehead. Night-time is for sleep, not pleasure. But a girl-child, for example, may prefer the earth under her fingernails, the aroma of manure and the touch of a dog's slobbery wet muzzle on her hand. She may grow up unaware she is considered more a pagan virago than a Virgin Mary and scorned, teased and rejected for her love of a more pantheistic lifestyle. This may manifest later as a sense of inferiority. Those who come close to her may find her controlling through her sexuality, or too moulded around her own beauty. She may look for partners she can submit to, or ones that she can manipulate then reject, before they get too close to her sense of inferiority.

Alternatively a woman may over-compensate if she is a very sensually aware individual, and if she hasn't received the same amount of exchange of awareness in body pleasure as she needs herself. She may become provocative, highly seductive, and over-tactile. Women are more likely to display this flirtatious tease than men because boys are socially conditioned, and expected to require less sensual attention. Boys are brought up to be potential winners, competitors, breadwinners and power tools in the West at least. Women still seem to haunt men's dreams as both mother and lover, offering a mother's breast and the devouring arms of her embrace. If our sense of love is triggered by our first experience of love, then our sensual responses or indifference may also be a repetition of our early erotic relationship to our parents. But this does not mean that it is our parents who are to blame for who we are, for we are already predisposed to certain qualities of being when we are born into this life.

The Witching Hour

Integration and acceptance of the opposing images of mother and lover can pose many problems for women today as it did during the Middle Ages. However, to be a sensual siren, exotic, independent, alluring and bewitching in the Middle Ages was likely to lead you to the stake, whereas to be maternal, functional, and wifely ensured at least that you followed in your husband's footsteps, rather than those of the inquisitor. It is the same dilemma now, but with a different twist. A woman's comforting, preserving nature may conflict with her desire and erotic pleasure-seeking nature; she may split the two apart and live out only one side or the other.

Many men were either witch-hunters or lyrical troubadours and knights – the roles they play now are similar personifications. Are they dominant control-freaks, plagued by power and routine, impassive and judgemental, or are they tender, sensually aware, receptive and creative? If a man finds it impossible to reconcile his image of the ideal wife with the sexual virago that has plagued his dreams for thousands of years as potentially his 'undoing', he may find it necessary to split these images too. How can one woman possibly be both to him, how can he experience the polarity without being reminded of both? He may turn to others in his search for sensual passion and bewitchment, for the experience of mother, wife and lover in one woman may be an impossible truth.

And what of those sensual ducking-stools? The witch-hunt frenzy like any shark attack leaves only fragments of tissue behind. There is only a red pool, a bloody ring of misty water spreading through the churning current where the kill has just taken place. But decay means growth, for every thread of the kill's body will feed another fish and others may live off the dead meat's blood.

The emerging arts of courtly love had a profound knock-on effect, being a turning-point in the history of our relationship to love and sensual pleasure. The early Greeks had wisely respected the power of Eros and sexuality, but the gap between nature and culture was now widening too fast. The terror of the witch-hunts was like an ice-bound sea. Nature had to find a breathing hole for the goodness and grace of earthly passion that was being steadily diverted and perverted into dogma rather than into growth. This breathing hole was found only in quiet places now. Romancing the sensual manifested in the courts, noblewomen's houses and the secret trysts, romances, art, missives and poetry of an emerging bohemian Renaissance. It was an intuitive sense of timing, the silken sheets only waiting for the final touch of those who lingered and loved.

Notes

1. Beaumont, cit Siberer, p.285
2. St Paul: Corinthians, 7, 1-2
3. cit. Young, pp.274-81

PART 5

Beyond the Looking Glass – The Value

'All that is vibrates with desire . . . the denial of desire will bring you only listlessness. Those who deny desire are the most smitten'.[1]

Beyond the Looking Glass

The true value of the sensual can be found in the realms of human relationships. In bearing each other, we have to respond to the senses our bodies share and feel, as well as from a sense of what the ego accepts, denies or that to which it attaches itself. In love relationships the energy of the sensual pervades even the ego, rocking its stability, shaking up its preconceived notions and expectations, always one step ahead of the mind's eye. A woman's sexual nature has been perceived as dangerous, and yet eternally in rhythm with the laws of the universe. A woman's capacity for sensual awareness is no different from a man's, but because of the historical distortions, most men have denied such awareness, and have seen it as only belonging to woman. Women then become what they are asked to be. They play the role of enchantress, vamp, Desdemona or Lolita because they have incarnated into social and family myths about who and what they ought to be, as well as being who they already are.

Men are as hungry for caresses, touches and embraces as women. Yet the irony is that woman has become an exaggeration of all that sensual quality of human love. The more we disown some trait in ourselves and fight against it, the more it is exaggerated in the other person. For example, the more a man disowns his tactile needs, the more it seems to be exaggerated in his partner. He may then begin to be repelled by her touching, her need for close physical contact. The sublimation of the sensual by Western culture over the past four thousand years has made many women schizoid about how beautiful, enchanting and alluring they ought to be, when at the same time, if they are overtly physical and/or sexual, then they are wanton harlots. It has emphasized men's fear that women are both enchantresses and whores. It is now only in relationships, particularly sexual relationships, that sensuality is cherished.

In romantic love it is the suggestion of the senses that evokes inner joy, first glances, touches and looks that 'set our hearts on fire', or sends shivers of desire down our backs. In intimate relationships our sexual senses, the most primitive and savage of all, are awakened. This is when romance becomes an erotic relationship, the need for the idea or the image to transform into something creative, tangible and mobile within ourselves. Human love, however, cannot sustain deep romantic-erotic attachment between two people for long. This is mysterious energy which works only through that which is in need of moving. Erotic love is dependent on the senses to keep it vital. But the senses are not necessarily dependent on our contemporary concept of love. The sensual prefers the force and energy of the unknown, desire must have something on which to feed itself, and Eros must bind and loosen, bind and loosen. There are as many different kinds of relationships as there are individuals which is why Eros flourishes in human company. We all have our different needs, values, feelings, psychological complexes and our unique presence on this earth and we all respond to love. The ancient Greeks had many different words for love, and many different ways of dealing with it, and several of these have become fused into an idealistic notion of 'love' in the Western world. This may embody the underlying pretence and illusion behind the investment we have put into one other person. How can all that we seek, all that we dream and all that we desire be constellated in one human being?

Chapter Twelve
Romancing the Senses

'Breaks passion, like glass shattered upon the senses'.

Romantic love begins with sensual suggestion. Romantic love is dependent upon the imagination, the chamber where Eros plays and lulls us into delightful dreams and sexual fantasy. The senses play with our feelings, aromatic messages laced with the heady bliss of ecstatic other-worldliness. We feel superhuman, immortalized, as if nothing can stand in our way. We are involving our senses in the imagination, tuning in to a mystical ecstasy known to many spiritual and religious people as enlightenment, or sensing the divine. An ineffable embrace that some find through the love or worship of a deity or vision, and most of us find in the eyes of another person.

Being in love lifts us out and away from our perception of reality, drawing us into ourselves and out of ourselves. But it is in fact a truly lonely world. These images and notions are not really the object of our desire, only the fantasy of our perceived image of that object. Once we 'get closer' to this object, whether through sexual encounter, or by companionship and friendship, we find that the illusion begins to fragment. It is dangerous, this being in love, but it is the most mysteriously enriching, rewarding and empowering force that permeates the human psyche, as long as we know how to be creative with it. Being in love with love means we don't want to ground the feelings, don't want to turn imagined longings into sensual pleasure and sexual union. Many people prefer the 'high' of the fantasy, rather than the reality of commitment to a body, with feelings, complexes and real flesh.

But romantic love usually transforms into sensual surrender. The shift into erotic pleasure, and the indulgence of sexual, emotional and ego sense however can bring pain. Then romantic love becomes the other, he becomes she and she becomes he. But the fairytale has no constancy. The truth of romantic love is that it is inconstant, ephemeral, a fabulous tale of impossible diversions. But romantic love is exactly what our Western culture has inherited as the most earth-shattering force between individuals into which we pour our body and soul. We cannot separate romantic love from companionship love, we mould them into a concept of wholeness and expect this wholeness to be found in one other person. Our demands are passionate, relentless and idealistic, even though our senses tell us otherwise. Romance becomes entangled with the body and sex. The heady, transcendent in-love-ness that we cry out for in the Other, becomes a real body, with feelings, sexual needs, toilet habits, bulges and emotional baggage. It is a feral world of vanity, envy and selfish devices, which now seems the antithesis of the heady illusory world of the first stages of romance. So how did we get into this pickle of love?

Idealization of the Feminine

Earthly love, or the sensual, has to find a way of surviving. Nature's cycles and rhythms often correspond to flowering times in history when adjustments, balance or chaos keep us shifting, keep the universe in relationship with itself. We are all dependent upon each other, the organism of the whole relies on each cell of itself.

In the medieval world there was Mary, there was God, there were witches, but where was earthly love? We don't really know how the ordinary people of this era lived or loved. The only written evidence is from those who were educated, religious scholars, nobles or artists, writers and poets. But the melody of the moment seemed to be needing a solid sense of a different kind of love. Mary was idealized mother and virgin, but in the courts of Europe, particularly France, there was a rediscovery of the beautiful woman and the love of her. It was timely to romance rather than be tortured, to celebrate the body rather than be celibate.

Around 900 AD culture was rapidly changing. A re-emergence of art and of beauty against the owl-light of the Dark Ages. Decadence was in decline and the spring of love was in the air. There was sense of something vivid, alive, eternal; something outside the compassion of Jesus's work and the fabricated absolutes of the Church. Style, colours, fashion and progress were alternatives for the witch-hunts of Church dogmatism and patriarchal rule. Medieval poets and singers sang of love and art; the idealization of a different kind of woman was spreading alongside Mary's compassionate 'other face' of woman.

The Lady of the courts was unattainable, only to be loved from afar, idolized for her virtue, her grace and her body. Like the Greeks, this new sense of beauty constructed a similar concept of slender body, milk-white skin, and suggestibility of what lay behind the clothing. Romance was about the unknown, about missives, messages, the lord's wife loved by the knight, passion in secret arbours and songs of love, suggestive, allegorical and mostly erotic. Courtly love was developing as a way to manifest Eros outside of marriage. As with the Greeks, as in many other cultures, marriage was a contract and love in all its guises was not necessary for it.

The archetypal nature of love, its evocation and its darkness still needed to find its place through human relationships. Eros moves into and out of us at will, and this force is unbinding, attaching, shattering and mostly transforming. It takes us into the darkness of ourselves, of our origins, of our battered winds, our storms, and our nightmare lands and passion. It is always this passion which saves us from the the mind's fear of death. Instinctual fear of death keeps us alive, but the human mind's conceptual fear of death keeps us fearful.

Courtly Love

Guilhelm, 9th Duke of Aquitaine, was one of the most powerful men in Western Christendom around 1100 AD. It was in courts such as his that the art of courtly love sheltered the erotic and sexual relationship between men and women. Guilhelm, a great hedonist, indulged in pleasure and sexual love with his ladies, his poets. When a preacher denounced the court antics as heretical, and began converting the noblewomen to the belief that hell was for all adulterers, men and women, Guilhelm changed his philosophy. Sexual love was no longer a sin against God, but a mysterious union between two people. The Lady became the embodiment of the divine, and woman was to be revered and loved as a goddess. Guilhelm was only one of many men who began to understand something other than Christian wholeness. He had found a way of synthesizing body, soul and the senses. Aphrodite had returned with a vengeance.

The first early courtly ideals of woman being placed on a spiritual pedestal became fused with pagan and Eastern sacred sexual practice. Love-desire was an ancient Arabic tradition whereby men would have to hunt and court the woman. The longer the hunt, the longer anticipation and desire were maintained, the better the love, the surer the ecstasy of every moment of wanting. Actual sex was irrelevant, although once captivated, the courtesan would be paid well for her services. The divine moment was also to be found through the sensual realms, techniques of love-making, probably derived from the Tantric arts of the East and brought to Europe from returning Crusaders, became popularized as esoteric spice. The synthesis of spiritual love and earthly love was founded on an ideal of woman as the source of grace and transcendence. The courtly lover became adept at sexual skills such as *drudaria* and *karezzam*, erotic pleasure based probably on the Maithuna of Tantra.[2] Like the Shakti consort of the Hindu gods, the lady of love was the creative process through which man could transcend body and ascend to soul. The profane was becoming a spiritual pathway again, as secret and as hushed as infidelity. Many of the songs were erotic allegories conveying hidden sexual messages that both knight and lady would understand. Women were also troubadours. One, Beatriz de Diaz, wrote:

> 'How I'd long to hold him pressed
> naked in my arms one night –
> if I could be his pillow once,
> would he not know the height of bliss . . .
> I am giving my heart, my love
> my mind, my life, my eyes.[3]

The fundamental need for courtly love had sprung from an era where roses had only thorns, and women's sexuality was the Church's peril. The beloved became an idealization of beauty, and like the hetaerae of Greece, the virtues, purity and pleasure of woman kept knights, jongleurs and poets returning again and again to honour her. A romance is a tale of chivalry, literature set in ivory and varnished with the jewels of love. The greatest romances, Tristan and Iseult, and the tale retold of Arthur and Guinevere from earlier Celtic pagan legend, became fashionable, and symbolic of the whole art of *courtoisie*.

Perpetual Fantasy

Romantics perpetuate the fantasy of perfection in another. What they are seeking is complete and utter ecstasy, fulfilment and a sense of the wonder of themselves through a relationship. Twentieth-century expectations and dreams are still constructed around a medieval culture that needed a counterpoint to the overture of piety.

Both men and women are tossed and driven by desire and the idealization from within. It has held firm, this romance with the pleasures of the mind and body, and does not give us much room for other imaginings. We often have an image of perfection, of an abstract ideal which another human being can never live up to. But we believe they do when we first meet, simply because we have not realized that the image we are asking for, the ideal we are crying out for, may be embodied only in part by this person and sometimes not at all.

The philosophy of courtly love relied on the fact that the knight never really got to know his lady, until that is, the sexual chemistry of a more earthly love pulled both under the bedclothes. Desire and yearning was the sense of love unrequited. Desire depends on the impossible, on expectation and the fantasy of sexual fulfilment rather than its reality. However, sex became tangled with this idealistic sense of longing, and romance encompassed not just the unrequited anticipation of passion, but the exchange of desire that can only become earthed through carnal knowledge. It was this carnal investment, the torn feelings, the moment of surrender and its fatal inconstancy, that bled Eros into the collective heart.

Infidelity

Much of the original philosophy behind courtly love centred around the unattainable lady of the court, wooed and adored by her knight. She would be married, not chaste, virtuous and unlikely to jeopardize herself in any way. Poems, whispers and letters were the only way these romantic night-stalkers could express their love. But there had to be a catch. Tristan and Iseult became the archetypal tale of doomed lovers and infidelity. Those faithful ladies of the courts became an image that women still try to live up to in their marriages. But were they so reliable, were they so true and loyal to their husbands? Iseult certainly wasn't, although she blamed the love potion for her passion. This is understandable, for every time we drink our own brew we are intoxicated by our own fantastic love of other and love of self. Falling in love makes us feel good, and if we feel good, as any addict knows, we stay with our addiction.

In medieval Europe love was freely given, not bound by the contract of marriage. Marriage was a business deal where desire played no part. Many other cultures, such as the Greeks, Hindus,

Chinese and South Americans, also contract-out marriage as a safe-haven for the evolution and continuation of the species. 'Being in love' was not the reason for marriage. Romance could be carried out in harems, boudoirs, Japanese tea rooms, secret rendezvous, or the sacred temple, but it was hardly conducive to the marriage bed. Nowadays in the West we believe that romantic love is exclusive to marriage and, more dangerously, that marriage is exclusive to romantic love, and it gets us into an awful lot of trouble.

Romantic Love and Epithemia

The ancient Greeks had a word for one kind of love, called *epithemia*, which was solely concerned with the instincts and basic bodily pleasure. We are all hungry to be touched and held, to ensure our world is safe rather than insecure from the moment of our birth. This fundamental human need became distorted with the belief that woman's sexuality and the body's fundamental needs were evil. What chance do we have to take pleasure in our bodies if we have a twisted view of ourselves? The joy of bodily touch, caresses, kisses and sexual enjoyment is a human need. *Epithemia* is also child-like, it relies solely on instinct and the need to be handled, as babies, to woo us out of our primitive state.

Epithemia was regarded by the Greeks as an indiscriminate way of taking pleasure, with no morals or judgement involved. It provided a safe container for the wild abandon of sexual or bodily excess, and offered a sanctuary for the guilt or remorse around bodily greed and instinctual hunger.

Taking pleasure in the bodily senses is one of the first notions that is suggested to us when we fall in love. Is this person going to be the most passionate divine person I have ever met? Will we make beautiful love forever, and can I experience the divine every time we fuse, or meet? But *epithemia* is an instinctive need, not an imagined bliss. *Epithemia* indulges the body, not the mind. This is where we confuse simple body urges with our emotional and abstract designs about love. *Epithemia* then gets confused with erotic love and particularly causes deep problems when involved with another kind of love, that of agape, the love of the gods for men, or a love without limits. Erotic love and *epithemia* may complement and enrich one another, but agape requires unconditional love, an ineffable, impersonal love which emanates from soul desire, but rarely one our egoic sense perception can even consider. Erotic love is transforming, passionate, intense and often sexual; it includes *epithemia* willingly, but is total immersion of two people with conditions usually attached. But agape is love that requires freedom; it may include sex, but not necessarily commitment to one other person. It says, 'If you love me you will let me be who I am without restriction. There are no limits. I am free to love another as much as I love you.' This is where the trouble lies.

Indiscriminate indulgence in the pleasure of our bodies happens all the time. We see someone, get turned on, fancy them, or feel we are in love. The problem occurs if sex is all we want or receive and the other requires more from us, or we find we ask more from them than they are willing or are capable of giving. Love becomes pricey. Separating our feelings from our bodies is for some an impossible task, simply because we have confused the unearthly imagined ideal with *epithemia*, our basic need for human contact and sex, and distorted both into an illusion of romance. We have combined all our senses into one word, that of love.

Sensual Openers

Romantic beginnings start with sensual messages, touches, caresses, looks, glances. Beginnings are about affirming and identifying that which will bring you pleasure, to both the eye, the mouth, the sexual organs and the feeling world. Suggestibility also gives us the first sense of whether we feel this divine shock living within. We may be bolted into the blue heat of love, like Semele's union with Zeus. Some of us prefer the romance of the quest, to be the hunted or the hunter. Sensual suggestivity becomes the tease of words, the provocation of nearness but never touch, and the voice on the phone can be the most provocative suggestion of what is to come. The voice is symbolic of the mouth, the lips, the tongue and the sexual kiss. Whatever our own sensual images are, we will find a voice that matches the notes of desire and the lyrics of our feeling world. In a sensual voice 'there is power, a strange power that beats upon the soul'.[4]

The way we dress ourselves, our beauty and our movement of body as we walk past our prey or the one who hunts us is an indicator too of what we are seeking to take pleasure in. 'I know too how the body should swing and balance from the waist and that is worth more in beauty than a slender line'.[5] Our bodily needs are often sublimated by the attitudes and mythology of our own inherited culture and social criterions. If we are aware of our body, we may understand that it is not separate from us, but part of us, working to take pleasure and give pleasure. Romancing our bodies means we can romance our lives fully towards the next stage of creative love, in the realms of sexual passion. The more subtle messages we become conscious of in ourselves, the more we may be able to understand what kind of sensual pleasure we truly need. The desire we feel envelop us, fill us and overpower us is a mystery, a force that is unstoppable, but if we listen to our body and watch the language of others, we may begin to discover that falling in love involves mind, body and, to the horror of many, the soul.

Romantic Futility

If we fall in love with love rather than with another, then we may find disillusionment and disappointment in the merging of our bodies. F. Scott Peck, in his book *The Road Less Travelled*, suggests that we cannot truly love until we have fallen out of love. It may be that it is not so much the falling out of love with the other, but we must fall out of love with the idea of 'being in love'. Falling in love with another can be a positive step towards knowing ourselves. To fall in love brings us closer to ourselves, for in the eyes of the other is the reflection of that which we are becoming. We see reflected in the other our own life-force, our purpose, our grace as well as our disgrace. When we fall in love we invest others with soul, our divine image which is mysteriously hidden from us, into which we expect the other to live and breathe life. We want this untouchable, unknowable soul to become embodied through the other's body. To discover the life-force and purposeful energy of oneself, one often has to find it first in the eyes of another person. To discover the beauty and inspirational quality in another, is to begin to discover one own's innate beauty.

Pleasure in the physical senses is part of erotic love, but not necessarily all of love. Pleasure and pain in the sense of awareness and inner wisdom is where the true value of sensually relating lies. Navigating love is like navigating with no stars to guide us. We have no map, only memories of other loves, recollections and fantasies, there is no 'real' sense, and we may need to learn to be aware of this senseless sense.

When the image we have invested into the other person no longer seems to correspond with that person, it is then we fall out of love. This is OK as long as we take back the image and know it as ourself. If we can reclaim our beauty as our own, if we can find through the pain of separation, sorrow, betrayal or emptiness a way of working with the soul's fragments, then love can move weave, transport and deliver. Love can become more than just romantic idealism, it can resonate through friendship, it can make us shudder with Eros, and like agape, the love of the gods for man, it can become an unconditional sense of love for all and other, rather than only self. Romancing the senses demands of us joy, but it also demands of us that we respect the power of the darkness. Sensual love can be both delightful and denied.

Sensual Potions

Doomed romantics like Tristan and Iseult drank a love potion, and later believed it was the strange brew which made them fall helplessly in love, denying all blame for Iseult's infidelity. When we drink the magic potion together, we fall victim to our enchanted water, infused as we are with an extraordinary awareness of all our senses being engaged. Look at Romeo and Juliet, Cathy and Heathcliffe and the lingering suggestibility behind *Casablanca*. The magic potion is more than just an ethereal chemical reaction, it is the chemistry of the senses.

The language of the body is closer to the language of the soul, because it cannot pretend to be anything other than what it is. So which senses get triggered first? Each of us responds in different ways. Usually sight is the first sense which initiates our headlong fall into the love potion and its bubbling essence of love. This instant attraction, as if fate had drawn you together, is both one's inner beauty seen as if in the other, as if reflected in a mirror, and also a mysterious and venerable force. This force has historically been personified in the guise of Aphrodite's son, Eros, a more potent and powerful an envoy than merely the boy of arrows and child-like penis.

So I see what I believe is the object of my desire, and he or she becomes the object of that desire. I invest my own sense of beauty, of love, of idealism, of soul into this desire object, but I have forgotten about the gap between me and you. The gap through which desire can fall.

It is only when we begin to engage the other senses that we begin to relate to the world of love in different ways. Smell is the most basic of animal senses. The scent we give off is ancient, earthly and often very sexual. Usually we get an unconscious whiff of whether this person will fit our image of who we want them to be. We sense them out without knowing we are doing so. For example, a man may deny his own receptive, emotional neediness and may unconsciously 'smell' out a woman who is covertly emotional, moody, needy and demanding, whereas on the surface she appears calm, pragmatic and bombproof. Women too are usually conditioned to buy into the myth that men are the rational constructor, the potent dominator in the relationship. She may deny her own potency, ambition and pride and 'smell' out a man who is covertly dynamic, aggressive, and competent when on the surface he appears quiet, artistic and doleful. Both relationships will be 'sniffing' out the others hidden qualities, because these are in fact their own hidden qualities which they are not living out.

But there are also many men who are deeply connected to the feeling world, who struggle with their supposed dominant roles and masculine assumptions. In romance we may find some men acutely vulnerable, and some women at their most powerful. Whatever 'smell' it is we respond to, it will always be a love potion peppered with a heady infusion of our own unconscious needs, values and qualities. By attending to the smells and reactions or responses to others, we may discover more about who we truly are.

In the Forest

When they drink the magic potion a couple are no longer part of the group, the party, the gang. They are separated from the world lost in the forest of feeling like Tristan and Iseult in the Forest of Morois. New senses sit uncomfortably in the forest clearings, the fog descends and the separateness of their loneliness becomes transcended. They believe they are one with the other, through the vital and most enriching embraces of sensual pleasure. If we deny ourselves sensual pleasure, if we run from Apollo like Daphne we are denying the power of nature that is in us. Romance takes us out of ourselves by escaping and unbinding us from our earth-sense through the illusion of cloud nine. But we must always return to the dark realms of the body, and remember we are no longer gods and goddesses, but earth-bound mortals and passionate darklings.

To stay in the Forest of Morois for ever, like Tristan and Iseult, is as unreal as the perception we have of eternal bliss. Our senses become too overloaded, sexual pleasure provides a meeting ground for confrontation, denial, accusation, jealousy or pain. *He* doesn't touch her in the right places, *she* doesn't respond to his kiss in the way she used to, or passion subdues or intensifies. Through sensual romanticism love is awakened, but love is not always kind.

Romance thrives on mystery, on the unknown. Romance swells if it is taboo, an impossible dangerous relationship, an unattainable wife, a man much younger, a woman much older, a step-father, a step-daughter. The essence of courtly love was that you must burn with desire for someone. Romance must be agonising, painful, mysterious and idealistic so that it could never be real. This is where the *sense of being* is neither pleasure or pain, it is both.

The dark side of our earthy sensual nature drives us into the most ruthless scenarios, we become wild women, wild men, uncompromizing, vengeful, lusting only our own desire to stay burning and painful. But romance cannot be sustained in marriage. Marriage does not pretend to be a container for nature's fury, because it is a cultural construction to protect the species against such chaos. It is we who pretend that we can find Eros's power and passion there, and we who foolishly believe we can control him. In courtly love, and in many of the forms of Greek love too, trivialization of the dark power of Eros was not the answer, but respect was. Maybe it is time to respect this force and our sensual integrity again.

Romantic Death

Capek's play *The Makropoulos Secret* concerns the story of a young opera singer, a woman who sells her soul to the devil. If she agrees never to fall in love she will have eternal life, She can have anything else, wealth, fame, power, beauty, but never fall in love. She is hugely talented, but empty inside because she can feel no desire. A young woman is willing to carry the curse for her, so she breaks her contract, knowing that now she will eventually die, but at least she will fall in love.

This story is profoundly symbolic of our romantic sense of life and death. The most fundamental instinctual need is that we must survive so that the species can survive. This is nature's tune, painful because we are conscious and can imagine our death, the difference between human and animal. Without the rich complex of light and darkness we do not feel alive when we fall in love. Our senses are barren, uncharged, and dead until someone awakens our senses within. Our egoic sense thrives, our physical senses are on red alert, and our feeling sense vaults any hurdle. We believe that 'fate drew us together', and it has meaning. We always feel something potently fated about such liaisons. Perhaps this is why like the opera singer in Capek's play, we can only feel vital and alive, burning and terrifying, erotically charged and ignited when we have entered into this natural chaos. Yet unconsciously we know we are mortal. Our bodily limitations mean there is a finiteness to our senses, so we must take the risk that fate plots out for us, enter through the gates of nature's ambivalence and leave the safety of civilization behind.

The medieval poet Montanhagol wrote, 'Desire never had any power over me to make me wish her to whom I have given myself aught that should be. I would not reckon that a pleasure which might debase her.'[6] Idealism means our imagination sets limits, we resist and kick against the terror of erotic desire and its power. To be seized by Eros was to be possessed by all the gods, for he brings his troupe for company. The courtly lovers acknowledged the mystery of erotic desire and attempted to find a safe container for it, but our contemporary backlash against anything which limits our freedom means Eros has become polarized between the sado-masochist and the white wedding dress. The awesome power of the erotic has been neatly fused, in our heads at least, with the simple needs of *epithemia*. But the erotic is more than *epithemia*, the erotic is a misunderstood force which civilization cannot contain. The erotic is a fellow intruder in our soul-house with the sensual. 'He' carries no morality, for Eros is a god born of chaos, not of order.

As she lies there, the first night of her honeymoon, does the bride surrender to the killing power of her desire, or once the white veil of cultural lace and illusion is removed, and the icing falls from the cake, is it she or her lover-husband who finds Eros was the motive for their desire or was it desire for the dress that aroused Eros?

NOTES

1. Roberts, p.63
2. Walker, p.862
3. Hill and Burgin, p.96
4. Fortune, D, p.35
5. Fortune, ibid
6. Briffault 3, p.489

Chapter Thirteen
Sensual Eroticism

Oh that I could be desire, could be that envoy of mystery to enfurl my cloak around you, to breathe desire upon you! But desire breathes on me.

The Body Again

We have an inherited value judgement about ourselves that we should not smell like people, and especially sexually aroused people. The taboos of Western culture have left us paranoid about our sensual experiences and the possibility that someone might 'sense' we are sexually inviting or have become aroused ourself. We struggle to find ways to disguise or enhance our pheromones, with deodorants, perfumes, body lotions and potions. However much we try to deny we are animals, we are still intimately connected to our ancient and humble roots of creation via our bodies. Yet when we are engaged in the most primitive embrace of all, that of the sexual kind, we are immersed in touch, smell, sound, sight and sometimes feelings, imagined or real. If we are honest, sex can be pleasurable without engaging in feeling love. The longstanding illusion for many is that we can find love, sex, companionship, structure, eroticism and romance all in one other person. And who can ever live up to those kinds of expectations? There is still, however, something mysterious and intangible at work when we indulge in the senses of sexuality, because it is the energy and power of this force which provides the interface between mystery and the body.

Symbolic Pleasure

There is more than simply man meets woman, or woman meets woman, or man meets man. There seems to be a deeper need in our relationships for us to find that which will make us whole, or a search for that which we perceive to be the quality we are lacking. We project ourselves everywhere. Every person we meet, or every group of people we can imagine, becomes a reflector for those qualities we may disown or have not acknowledged in ourselves – both beautiful and ugly. It is human nature to do so. Without projection we would not be able relate to anyone else. We could not bear ourselves.

They say opposites attract, and worn-out clichés are sometimes invaluable because they usually hold a profound truth. Whatever our sexual preferences, we still seek that which we believe we do not own, in order to resolve the illusory dilemma of the 'missing half', or the soul-mate. This is nature's creative trick, for with this urge to unite, nature carries on creating life and the cycle repeats itself. The life-force and the conserver of that life-force are within each of us, but we usually deny one of these energies and look for it elsewhere.

Spiders are symbols of women, as women are symbols of the fruit of life, and the fruit is the symbol of the origins of all creation and potentiality. Woman entices, spiders set traps with their webs, the fruit entices us to eat it with its juices and colours, and sensual messages abound. A woman is enticing the man who must eat her. For only through the eating of the fruit do we acquire wisdom and creativity to honour the life-force and soul. Through a woman's body men are held in the boughs of the tree of knowledge. Through a man's body, a woman spins the mystery on her web, and together they eroticize and mobilize the life-force.

The Stranger

The initial process of desire is as much a woman's dilemma too, because she unconsciously knows she represents to men exactly what he most fears. 'Fated attractions', 'fatal attractions', 'it was fate that threw us together', these are all phrases and epithets for the initial process of desire. The stranger is a mysterious suggestion of what is to come. Intimacy brings with it familiarity, not an easy zone in which erotic love can breed. The first few months, few years even, we may believe we are set for a life-time with a 'stranger', but when the visors are lifted from our faces, we find only another human being, not an imagined perfect other, nor an unknown stranger.

The 'chemistry', pheromones flowing, eyes gazing, body teasing, are all ancient messages that cause us to engage the senses totally at risk. Haphazardly we fall in love with someone who may be nothing like us or may share nothing in common with us, simply because we are not aware that those qualities we have projected onto, or desire from the other person, are in fact those very qualities in us which we may have disowned. Desire is Eros's *agent provocateur*, and it both serves our biological needs as a species and provides a powerful source of psychological growth, so we can at least begin to reflect upon ourselves with consciousness.

It may be a manifestation of civilization that a woman symbolizes man's greatest instinctual and philosophical fear – that of the mortality of humanness – but it seems *she needs* to be this symbol. A woman has to preserve and nurture through her sexuality, her earthiness, her rhythmical finiteness that is nature's demand. A woman often reminds a man how terrifying, how awesome, how wild and unpredictable nature is, and also how he has sold out to instinct and bought into synthetic living.

Men and women still confuse erotic passion with bodily need, and sensual pleasure with love. The fusion of the sensual with sexuality, the fusion of romance with erotic love, are what make us so human and sometimes crazy. Mankind has unconsciously asked woman to carry the quality of 'sensuality' as part of their sexual nature so that man and woman must fall into each other again and again. The mysteriousness of life has locked itself away in men and women so that we may keep finding it unexpectedly in those moments of chance encounters, first kisses and the chemistry between woman and man. Plotinus said, 'soul puts lures in things' by which we will be drawn to find the mystery of life and its meaning.

Lonely Self

Every person is an alone self. The world revolves around us as babies and children, and we grow up tossed and turned by the winds of the family fable, family curses or seemingly outer events which come to shake up our lives. But we are as much part of this outer world as we believe we are isolated from it. We are ego-centred and ego-driven creatures who perceive and imagine the world around them totally subjectively. We are all drivers of our own vehicles, and yet we are also the car, the back-seat driver, the road, the pedestrians and the traffic lights. Relating through the senses is instinctive. Bodily pleasure is simple, it carries no judgement, no intention other than to please itself, and possibly the other. But the ego feeds upon this body, and it may revolt against the instinctive nature of our carnal beginnings. Relating through the cultural arbours of emotions, feelings and social expectations we are dazed and then delivered with distorted idealizations. We enter relationships through the fantasy threshold, our complexes and psychological dressings collude to deny our basic needs and sublimate our cravings or our fears. So lonely self looks for other lonely self and assumes it will find home.

As children and babies we are programmed immediately into a sensual world. It is our survival mechanism to suckle and to be caressed. We must involve our senses to grow and develop into lonely selves. However, no lonely self can ever grow up uncontaminated by society, family, and their own psychological motivations. Without your mother complex or your father complex you cannot have any kind of relationship with the world. It is precisely these outer dynamics that play into the psyche that can be richly rewarding and creative in relationships if we attempt to make these shadows of strangers our conscious friends.

Mankind's fear of the ambiguity of nature, as we have seen, has lead to a need to contain it, or lose oneself to the mind, a far safer place than bestial nature. Every time a man looks at a woman, every time he sees his unhealed wound he is reminded of the knife that slipped in Chaos's hand. Her presence carries powerful reminders of the chaos of nature. Many scholars argue that the dichotomy of man versus nature is a modern cultural construct. Whether it is a myth we are buying into at the present time or not, the weave of humanity's loom has reached a point in time when we are making conscious the possibility that whether this is a true split or one in our collective imagination, either way it is in need of attention. Every myth holds a deeper truth, like a doll inside a doll inside a doll, as you reach down into the next container and open the doll, the truth does not lie in any one doll, but in all the dolls and what you consider the doll to be.

We are currently at the threshold of a new astronomical age and there seems to be a conscious need to redress the balance of the feminine and nature. For it to be re-established as of value and of as rich and deep a meaning as the masculine. However, if women chose power and yang-assertiveness over the deeper, gentler, mysterious and sensual qualities of femininity, it may mean we lose that very precarious disorder of the feminine symbol itself, the chaos which forces nature to be creative.

Women as fundamental symbols of the feminine, whether as nature's duplicity, or nature's diversity, may need to review the sensual, provocative, enticing qualities of human nature, just as vividly as men. Not as compensatory action for what has happened in the past or their own personal issues, but to treasure and restore the place where humanity may find the profound truth about itself, through sensual awareness.

So Who Is This Eros?

'Desire' is deeply enmeshed in Eros's primal base nature. Being driven by desire is to navigate with no stars, no map, no logic or human rationality. It is to be overpowered by the god who seeks only to transform and unite the opposites.

Eros is an ancient, raw, instinctual, archetypal energy. He represents the urge to unity, and has come to drive us towards acknowledging the base nature of ourselves. Eros does this through transformation in relationships, for he is not a god of civilization, of knowledge, nor even truly of the Olympian pantheon. He pre-dates history, both as a god and the energy he symbolizes. He is both creative and destructive in one force. This is the feeling or the compulsion of desire. Eros is the giver of desire, but also ultimately the taker of it.

Eros is similar to Kama, the early Vedic god born from initial creation. This primal force is the dynamic masculine principle which was a necessary balance to the receptive, creative aspect of the Shakti consorts, or in Eros's case his ancient mother Chaos. Eros is also symbolic of the moment of union in the creation of a human being, our own parents being the vessel through which this creative life-force became fused. As soon as we become a fusion of opposites, of yin and yang, of male and female cells (whatever gender messages we genetically receive) we are also immediately isolated and separated from the mysterious source. We are suddenly no longer at one with whatever it is we are mysteriously created from.

Eros represents our attachment to the life-force and it is often sexuality that is its most powerful expression. Eros binds us to others, he also transmutes the relationship, twists and shifts it when necessary. He may also loosen our grip on reality when we become obsessed by the urgency of desire. The erotic is not simply the porn shop and the sex film, it is one of the most powerful energies of the universe, symbolized by the potency of women's sexuality. To many men, this is one of the most alarming aspect of Eros's excellence.

Female Sexual Power

The complex psychological nature of women is not so very different from that of men, except that woman has come to represent all that is feminine, both dark and light. With inherited values about how we should behave as men or women, we wrestle in confusion about whether we should seem more feminine or seem more masculine. Many men have cast off their feminine skin-sense like a regenerating snake and projected this side of their nature as woman's sole responsibility. But Eros takes us all into the complex nature of sexual relationships – Aphrodite's realm. Aphrodite is the pleasure-giver and taker, while Eros is her envoy of desire. Fusing Eros and Aphrodite as a feminine embodiment of love means erotic love preserves and perpetuates the force of regeneration and transformation, rather than merely conquering it.

Take the rich, dark and ruthlessly bitter tale of Phaedra. In Euripedes' play is revealed the dichotomy of humanness, and particularly an archaic perception of femaleness. In Phaedra herself we see symbolized the inadequacy of culture when faced with the relentlessness of sensual irrationality, in which women apparently flounder or delight.

Female sexuality was considered both inspiring and dangerous by Greek writers and thinkers, and out of necessity they chose to contain it. Nature's eroticism is ambivalent, and Western eyes are trained to prefer the axiomatic. When Eros breaks into our lives, as he often does, we are ashamed, confused, guilt-ridden or obsessed as we toss and turn under his spell. A woman may find her own sexual and sensual needs alarming, unusually active and potent. She may do anything to meet her lover, lie, pretend, deceive, cheat and deny those feelings within herself, simply because such passion is alien to her more familiar sense of social expectation. A man also may find the power of Eros uncontainable. He may weep for release, be torn by desire, wracked by guilt or feel bound and suffocated, jealous and driven to push further, to look down into the abyss of his darkness. She or he has met the outer limits of human experience at the edge of the underworld bypass.

Phaedra is the female protagonist in Aphrodite's story of revenge. Briefly, Phaedra's stepson, Hippolytus, refuses to worship Aphrodite, prefering Artemis the chaste huntress. In his worship he has vowed celibacy. By definition human beings are sexual animals, our bestial instinct merely dressed with the trappings of consciousness. Aphrodite, vengeful and spiteful to those who do not honour her, seeks to destroy Hippolytus and demonstrate her potency, for to deny one's sexuality is also to deny one is a human being. Hippolytus's arrogance must be punished. Aphrodite curses Phaedra by making her fall passionately in love with Hippolytus. 'When Love sweeps on you in her full power, to resist is perilous', the Nurse warns Phaedra, who cannot resist. Phaedra is married to Hippolytus's father, Theseus. In the throes of insatiable desire for her stepson, she cannot resist revealing to Hippolytus her love for him. Hippolytus, outraged, rejects her, and then proceeds to give a monologue on the evils of women, especially those with a brain: 'The sexual urge breeds wickedness more readily in clever women; while the incompetent are saved from wantoness by lack of wit.'[1]

Before Phaedra hangs herself in despair she accuses Hippolytus of raping her in a suicide note, carefully left for Theseus to discover. Banished by his father, Hippolytus receives his come-uppance as he drives his chariot along the cliff-tops. Poseidon sends a savage bull to scare the horses and the chariot crashes on to the rocks below along with the still-celibate Hippolytus. The forces of nature, and the savage chaos of sexual love that Hippolytus so desperately denied in himself, became in the end his fate. The power of woman's sexuality symbolizes the archaic archetypal energies that evoke such passion. If we do not honour the fact we are human, whether our gender is male or female makes no difference, we may be in danger of either believing we are omnipotent, or being forced to confront our own fate.

The New Phaedra?

Passion does not change, because passion is an energy that defies civilization. Nor does Phaedra change. She is like many women, who become the protagonist for a man's unconscious daemons. Recognition of one's sexuality means one does not deny one's humanness.

Take an example of a new Phaedra. She may fall desperately in love with the wrong man, at the wrong time or in the wrong place. She may say, 'How can you be married and have three children? I want you, I'm passionately in love with you, give up your wife, give up your home for me!' His marriage to someone else, in a sense like Hippolytus's marriage to his religion, means exclusion of all else. So he may give her some hope on one level, because when faced with passion and mystery what human being can refuse? If he believes he will stay true to his marriage in 'spirit' and in mind, not in body, then he may pretend to himself that he is only being a man. Carnal lust is the guilty party, and his 'sense' of guilt denied. Passion becomes the excuse for his actions, not his ego's intention. This new Hippolytus may enter the relationship with expectations only of sexual reward. The original Hippolytus was not tempted, but his very vexation and rejection of Phaedra and his eventual fate are still an impassioned response.

New Hippolytus replies, 'You know I'm crazy about you, don't you? Soon, I'll give up my wife, I promise you. Just give me time . . . You know how I just go wild thinking about your body?' The new Phaedra now has her way, spends a few romantic dinners, a few stolen afternoons, a few lost weekends. She will send him secret missives to his office, get friendly with his secretary and hang around street corners waiting to catch a glimpse of him as he sweeps off to the gym, or on holiday with his kids in tow, his wife hopefully nagging him. If Phaedra sees him kissing the wife she may start to phone late at night, pretend to be his associate, or hang up in tears if wife answers the phone. Then comes the resentment, the jealousy.

'Why didn't you answer the phone last night?' she demands as they roll apart from bodily pleasuring. 'I need to talk to you at three in the morning, it's important, I can't live without you! I may do something insane!'

'Hey, I'm trying to sort this thing out, give me time, give me space. Don't get possessive, and don't ring me at home. It's dangerous, she may find out, and what about the kids? Do you want me to hurt my family?'

He will always be torn apart by his betrayal. Hungered and seduced by his own desire, he finds the words flow as an easy option to subdue his lover. Little love, however, may be invested in Phaedra. Oh, he may admit to being in love, to undying desire, to romantic passion which takes him away from the routines and responsibilities of life and contracts of family and marriage. He may give in and give up his life for another woman, for passion is as twisted as life. But if he doesn't, our new Phaedra will find tears and pain, the longing to die, the longing to seek revenge, the longing to leave a suicide note and hold him responsible. For the new Phaedra may expect love and emotional response, complete and utter intensity and commitment from her lover, not just sexual pleasure. She is confused, possessed and possessive, painfully engaged in erotic love which thrives and grows as rapidly as a sequoia tree when watered with sensual reward. We do not relinquish the desire so easily, nor does desire relinquish us.

Betrayal

Engagement of feeling and erotic pain has been perceived as a feminine principle, and one which men find the most hateful or fearful sense of themselves, and of women too. Men too can be hurt and betrayed by women, for we are all human and all vulnerable. But mankind has made its own *soup du jour*, by putting women in the cooking pot and making men the ones who must eat it. By the time Western civilization was embarking on a complete revolution in communication and exchange of ideas in the Renaissance, once the printing press was churning out books to those who hungered for knowledge, it was not long before man's perception of women as temptresses, *femme fatales*, vixens and Liliths became fixed images in society. From the Renaissance onwards, and its breakthrough in wisdom, science, knowledge and discovery, women have been perceived as a sensual jinx, wicked yet frighteningly irresistible. Similarly, it is hard for us too to separate the senses from the ideal, the pleasure from the commitment of lasting bonds. If we have both, we are blessed, not lucky, but it requires work, awareness and few illusions.

Monogamy or Monotony?

The neo-Phaedra is often to women what Theseus is to men. Phaedra becomes bound to her sensual passion through her feelings, Theseus attempts to attach himself to the abstract world and detach himself from his feelings. But if neo-Phaedra enters into a long-term relationship can she sustain the sensual mystery that she once found in a stranger?

Here is another example of a neo-Phaedra. She may find great security and a long-lasting bond between herself and her partner. This is the power of a contract. But the power of Eros is not about contracts it is about attaching to passion, to both the binding and the unbinding force of transformation. Her relationship may become dull and routine, it will be static, no longer volatile and emotive, her charm and beguilement possibly redundant. She may fantasize about other men, still dress to attract and flirt, to kill with her looks and magnetize with her sexual power outside of the efficient engine-room of the marriage. She may do this oblivious to her own motives, but if she does it, it is because she seeks Eros elsewhere.

The scenario might go something like this:

'He's so boring, all he does go to the pub and talk about TV. I want excitement, romance, passion, like it was when we first met. Not nappies, housework and routine. Maybe I should take up my career again, something has to change.'

'Go and buy that book about how to improve your sex life.' suggests a kind friend. 'Or try something kinky or naughty at that new sex shop.'

'It's not the sex really. I mean we enjoy each other's bodies, and we probably make love three times a week, which I've heard isn't below average for most couples – but I don't know, something's missing.'

'Come to Gemma's party on Saturday. Go on, have a laugh, flirt a bit with all those wonderful city slickers, it won't do any harm?'

'I don't know. I might feel guilty. Bob's going on a golfing weekend, and I really . . .'

So she goes. Two weeks later in the same coffee bar, the same friend.

'God, I think I've fallen in love,' she whispers. Her hair is shimmering, her make-up immaculate, her body moving with grace.

'Not thingy, Emily's partner? Hey, you could be getting into a dangerous liaison with that guy.'

'I don't care. It's beautiful, the feeling is beautiful. I want to be in love. I don't care about what happens. Bob won't ever know about this. Just once, I just want to experience that moment again, just once!'

Phaedra dresses to kill, adorns herself, secretly meets thingy, falls into bed and falls into love, and she is doomed in some ways. But through her dangerous liaison her senses come alive again, not just sex with a partner, but sex and terrible desire for a stranger. She is the court lady, she is Phaedra in blind passionate love, she is frozen in time by her desire, knowing it will shift, for it always shifts, and moves into another dimension. The knight of her dreams may idealize her and fall in love too, but the modern dance has a twist to the ancient steps, for unlike the ancient Phaedra she may expect all of it – the passion, the union, the completion, the bond, and then the divorce. It will not be a suicide of body, but a suicide of love. The neo-Phaedra may well evoke need as dangerous relationship, drawing to her those who will embody the self-sabotaging aspect of herself.

The modern Theseus, on the other hand, usually avoids any painful commitment to Eros. He may have been a womaniser when younger, but now is resigned to a marriage where he can have his cake and eat it. There are many men who have deeply erotic and sensually loving relationships, but how many of these are doing so with women who have been their wives for a long time? The relationship may have begun in this way, but it takes a very unique relationship for erotic love to constellate perpetually. Eros appears to live through the mystery of woman which many men so desperately long to make their own. Eros moves just as mysteriously through men as he does through women, but Eros has to side with Aphrodite, he has to be embodied in the feminine image otherwise he becomes vulgarized and cheapened, a porn video or an S & M freak.

Theseus spends more and more time at the golf club, but what about Eros? Eros always finds his way back into our lives if there's the slightest chink in the curtains.

Theseus might have said to Phaedra his wife, 'Golf this weekend darling. You remember, the boys organized it weeks ago? See you Monday, enjoy that party on Saturday night at Gemma's.'

Three hours later he is on his way to London. At a pavement café he meets Jane. The pigeons are squatting, strutting, pecking, tourists necking.

'Was it difficult, I mean, getting away from your wife? Lying about the weekend?'

'No. She's got no idea. Gone to a party.'

'I'm really nervous about this. I mean, what if . . .'

He touches her lips with one finger. 'This is our weekend, no one else enters the room, just you and I, just pleasure, just us. OK?'

Unconditional Intimacy

So what are we looking for in our most intimate relationships if it is not love? Could it be we are also hungry for mystery, in itself one of the aspects of love? We are looking for excitement, romance, we are looking for the stranger who will reflect our image without distorting what we expect to see. Once Eros has caressed us with his awesome fingers we come down to earth again with a bump when we find the shadow in the mirror is ours, and not that of the dark stranger's. Familiarity is not erotic, the exotic and the unfamiliar is. Predictability allows little room for transformation, but many of us long for intimate monogamous relationships that will be filled with Eros forever.

We even believe we can keep an erotic relationship alive by working hard at it, changing our sex manuals, swopping partners or indulging in sex toys and erotica. Like Phaedra, to be smitten by love and desire is the most initiating and yet odious moment when love finds us already smitten with a routinely caring relationship with Eros washed out of its socks. The stranger will be empowering, and our senses stir into action. Awakened to the new face in the mirror, we project different qualities onto this person, or perhaps simply re-invent the same qualities we saw before. The stranger who awakens desire is a mystery. It is mystery which brings passion, it is mystery which brings pain and pleasure, and it is mystery that has become symbolized by woman. The desire to merge souls, and the state of being in love jostle with the blatant carnality of sensual pleasure. Sensual experience does not rely on love, love however relies on sensual experience. We have a sense of what love does to us, we have a sneaky sense of what our bodies want, and we have a fearful sense of those intruders, envy, hate, jealousy, vulnerability and self-destructive emotion. We want it all, we want passion and love without the pain, we want to desensualize ourselves from feeling anything except wonder at the mystery.

Kissing the Mirror

There is a basic animal need to be touched, for the health and survival of the species. This does not necessarily only mean sexual contact. Grooming, sensual affection between mother and child and the interweave of sexual suggestion are all important animal needs and instincts for survival. Animals are innately sexual, and so are humans. However, social genres have imposed a distorted view of our bodily needs, and so we often suppress or sublimate them. The repressed sexuality of women can manifest itself in any form, as frigidity, workaholism, earth mother, prostitute or promiscuity. Denial of sex in men may result in celibacy, addiction, power, obsessional sexual addiction or sadistic or perverted sexual behaviour.

Kissing is probably one of the most basic sensual messages we convey to another person. We do it with friends on cheeks, kiss the air if we are high powered career people, and gently plant our lips on babies or an aged parent's forehead. Every society has its own rules and conventions about kissing. The Russians, and many Middle-Eastern cultures, are very open about kissing. Men kiss men on the lips and in Egypt men touch men, women touch women, but women can't touch men in public. Similarly, people from the Far East and Middle East are often shocked when they see men and women touching and fondling each other in public in the West.

In intimate relationships, the first kiss is probably the most erotically charged, the most suggestive and the most intentional message of desire. Our mouths are sensory temples, our tongues, lips, inner cheeks, and salivary glands evoke pleasure. Babies use suckling to survive, we eat, speak, grin, smile, suck, nibble and taste with our mouth. We convey more messages than just spoken language. It is the realm of the most sensual and erotic first encounter in our search for self in the other.

The raw, instinctive nature of men and women was one of D.H. Lawrence's most passionate themes, 'She took him in the kiss, hard her kiss seized upon him, hard and fierce and burning corrosive as the moonlight. She seemed to be destroying him. He was reeling, summoning all his strength to keep his kiss upon her, to keep himself in the kiss.'[2]

When we 'keep ourselves in the kiss' we are delaying the cycle of growth and decay, for as soon as we begin a relationship, we are beginning a journey which has no turning back. The only way out of this kiss is through pain, transformation or the kiss of another for 'when you have survived the terror, the first glances, the longing at the window, and the first walk together, once only, through the garden: lovers *are* you the same?'[3]

The kiss is both a celebration of mutual desire, and a moment when we are doomed, as if we have tasted the love potion. Once we have joined lips and let the surge of sexual suggestion forge through the body this kiss becomes the most potent intention of our needs and feelings. Yet sadly, as the relationship moves, changes, keeps seeking other avenues and streets for its dance, the kiss may no longer hold the very charge of mystery and desire that is the essence of that first moment our lips unite. In a sense, when we kiss another's lips we are also kissing our own.

Reflecting Whose Image?

Victorian poets, writers and artists became fascinated by the belief that a woman was so self-involved that she spent most of her time kissing her image in the mirror. This ancient auto-erotic image was to become one of the early nineteenth-century psychological models for narcissism, born out of the myth of Echo and Narcissus. A woman's cyclic nature was also perceived as mantric, personified by Echo who could only repeat herself eternally. Narcissus the effeminate youth, chained to the reflection of his own beauty in the pool, was regarded by the Victorians as the autonomous self-love of woman, gazing longingly at her face and likely to find the pleasure of self-love more potent than men's.

The mirror became one of humanity's symbols of woman's reflective nature, her moonlit nocturnal haunting; of dark nights, of oval grim mirrors reflecting shadows cast by the howls of wolves, sirens and enchantresses from Circe to Christine Keeler. The sensual creatures that roam in search of men – greedy werewolves, fox-women or vampires of love ready to drain men's life-force – were as prevalent a myth across the Far East and Africa as they were in the

West. In the culture of the late nineteenth century it is almost certain that many women did prefer love of themselves and each other to this revivalist misogyny.

Zola's Nana takes great pleasure in 'gazing at her reflection . . . oblivious of everything around her'. 'A passion for her body, an ecstatic admiration of her satin skin and the supple lines of her figure, kept her serious, attentive and absorbed in her love of herself.'[4]

This powerful image of self-love may be a literary manifestation of the collective genre, yet it holds a profound truth about relationships. How we perceive ourselves in the mirror is often how we perceive the Other who plays out that part of us for us. For example, you may see your reflected image as blemished, too fat, too thin, dulled by work, sullen and ugly. This may be how you relate to a partner after the initial mystery has been polished off the looking glass. He or she may be too thin, too fat, you criticize and nit-pick, you complain and think he or she is dull, bitter and works too hard. Alternatively you may see yourself in the mirror as beautiful, refined, artistic, dedicated and charismatic. You may see your partner in an equal light, always perfect, enchanting and faultless. But what you are seeing in the Other is exactly what you are seeing in the mirror as your perception of yourself. You may not only imagine it in your partner, but you may also manifest it in your partner if you do not own it for yourself.

It may be worth asking oneself these questions: What is it we ask the mirror image to be? What is it we ask or cry out for in a partner and ask them to be for us? What is it we are imagining this image to be for us, or this person to be? To be driven by desire for the image of oneself in the mirror, is no different from the looking glass in someone else's eyes.

The Victorian obsession with woman and her looking glass was a mirroring of the growing need for society to look at its own reflection. Up until the end of the nineteenth century, women had always been seen only in relation to men. Rather than being merely the reflections of men, women were now seeking men to be reflections of themselves. The turn of the century intellectuals, artists and writers, as well as the parlours of polite society, were faced with the new autonomous woman, in no need of dying for love of her man. Her own moonlit world was becoming fearsomely uncompromising.

The Power of the Kiss

The kiss caresses us whether we place it upon the mirrored image of our own lips, or on our partner's lips. To kiss is to receive that which we are giving, it evokes a passive power, a yielding and a possession; like the waves or the moon, it waxes and wanes, it is tumultuous and it is abandonment. The kiss is a human art, and yet for many men in the West, it is perceived as their curse, and their powerlessness in the embrace of a woman. *She* becomes stronger in his kiss as *he* becomes more vulnerable in hers. The exchange of energies is the sacred fusion of two bodies before the final embrace. The kiss symbolizes the moment of creation, of that which must come into being if it is to be. Both are strangers, creation and the moment. Creation is the misfit, for desire will fade as the moment moves. Infinitely unknown the first kiss is the last kiss. For this is the moment of sensual pleasure at its most profound. Sexual embrace becomes the heart, coitus means literally 'to travel together', but the first kiss is unknown terrain, we do not travel together in our first kiss, we travel alone. The first kiss is the stranger to the heart. It is unpredictable, it is not heartfelt, it is unconditional passion, a moment that can mean all or nothing. The kiss is the beginning of the journey, both the waterfall and the pool for those who dive or dare. 'Let him kiss me with the kisses of his mouth; for thy love is better than wine . . .'[5]

The moment of the first sensual awakening in oneself and in the other is an erotic conspiracy to fuse the opposites, to de-polarize that which is separate. The very nature of Eros is protean, and it is exactly that process of transformation which most of us find impossible to reconcile, for how can the first kiss ever be repeated again with the same partner? Disillusionment offers us the chance to find true sensual grace and creative life, unless of course we are happy to keep falling in love, keep searching for the first kiss, keep perpetuating the myth. Idealism and disillusionment are the two most dangerous of sense perceptions, but also can be the most enriching qualities towards finding any meaning, mystery or enchantment in life.

The Embrace

If the kiss is our taster of a more sensual and emotional feast to come, then the embrace of intimacy absorbs us into the world of all the senses. The senses beyond the ego's bounds.

> 'It didn't feel like it was generated from my own egoic sense. It was more of an etheric sense, it was outside of my body, surrounding my body. It was an energy movement. It felt like a presence, a sense in which this play was happening. This kind of archetype of love, symbol of love, was happening within this presence, this very still, rested presence.'[6]

The Taoists, the Eastern mystics, and practioners of sexual alchemy including Tantra, all revered men and women as possessing a powerful sensual connection to nature. The Tao is embodied in that which flows and changes, it is the moment in every moment, the synchronicity in everything. To resist change is to resist life. The female body was perceived as poetical, the words and art of the ancient Eastern cultures looked upon women's sensuality as auspicious. To lie with a woman would improve man's own potency, and the retention of semen was a potent part of the act of love-making to promote longevity, or some believed, even immortality. The release of sexual energy and the interaction of male and female bodies in slow, and often exaggerated love-making not only improved one's sense of harmony and sexuality, but improved one's connection to the divine. It was and still is a way of being. There is no problem with sexuality itself, it is our attachment to it, our ego's binding to sex that is.

Erotic initiation led to mystical union, and sexuality was sacred, as it had always been in the ancient civilizations of the Middle East and much of the pagan West. Pagan practice of sacred sexual acts of worship continued throughout history long after Christianity's rise, but because in the West woman was considered the source of all evil, most of the mysteries, esoteric and erotic sacred ritual and worship vanished underground, in a secret embrace of its own.

The world of sensual passion is alien yet compelling, it is throttling, duplicitous, amoral, anti-social, obsessive and de-stabilising. It appears to go against civilisation, and thus goes against mankind's phallic fallacy. Eros seeks to break and bind, break and bind, to fool us finally into Maya's arms. So who is Maya?

NOTES

1. *Euripedes*, ed. Radice, p.96
2. Lawrence, p.81
3. Rilke, p.10
4. cit. Dijkstra, p.136
5. *Song of Solomon*
6. Quoted in Ramsey, p.181

CHAPTER FOURTEEN

Why Men Need Seducing

'Women must carry it all: feeling, fantasy, loving, relating, inspiration and initiative'.[1]

What Is Seduction?

To seduce literally means to lead aside, or lead astray. The act of seducing is to lead someone out from themselves, to discover the magic and illusion of Maya through the spell of enchantment. In another sense, it is to lead one away from one's subjectivity.

Maya is an ancient Indian enchantress, an illusionist, a sorceress supreme. She enchants us into life. She was identified as one of Kali's emanations and Shakti, the Hindu creative feminine force. She is both the female form of our perceived reality, and the principle behind that reality. Maya is the web of our being and is the creator of earthly appearance, which like any appearance can be deceptive. In Maya is embodied the acceptance of tangibility but also the enigma of this illusion.

The dualistic view of the world is an illusion, this is Maya. We in the West are often conditioned to process the world in separateness, to distinguish between this and that, him and her, me and you. Most Eastern cultures and many modern mystics, artists and writers consider reality to be unbounded, that we are all connected as one. 'The East, in seeing that reality was non-dual, not-two, saw that all boundaries were illusory.'[2] When the magician performs his illusion on stage we know it cannot be real, we know the woman cannot be really cut in half, we know it is a trick. Like the illusionist we have created similar boundaries around what is and what isn't. We have a distorted perception of something we name reality, but reality is like the bisected woman, a trick of the mind. This is the illusory world of our perception, and each of us views the world from a subjective viewpoint. Each of us embodies Maya.

A man may be suspicious of a woman because he believes she represents everything that is illusory and mysterious. It may be that this man is not aware of his own mystery or is fearful of the unknown and projects it onto every woman he meets. Equally, he may desire a woman because she evokes the magical in him, or because she resembles the image he has of the sculptor or magician of the illusion; she becomes Maya herself. The one mystery we are all hungry for is Maya's mystery, the secret of the seamless coat of the universe that we participate in as a Whole. 'We are the bees of the invisible . . . we deliriously gather the honey of the visible, to accumulate it in the great golden hive of the invisible.'[3]

Is is through sensual seduction that we learn to look outside and inside of ourselves, to become more objective and aware of who we are by integrating those qualities in others to whom we are drawn.

The Mystery of Seduction

It seems we may begin to discover this mystery when a woman seduces a man through her senses and feeling world. The body of woman absorbs the out-breath of eternity. Her sexuality is both the gateway to human earthly pleasure and the threshold to life. Her ambiguity lures and attracts as the vessel for life. If she can carry a child, she can also carry a mystery. She must be receptive as a man must be potent.

One of humanity's universal seductions was that of the gods through worship and veneration. To venerate means to revere, but it also meant 'to desire'. The now archaic word 'venery', means sexual desire. One source suggests that to venerate is 'primitively signified as the sacral act of alluring or enticing something from beyond mankind's power'.

Seduction depends on the seducer and the seduced. The defensive boundary of separateness as an individual, becomes no longer separate when seduction moves mysteriously between two people. We are called individuals, and believe we are individual, separate, unique, alone, which to some extent we are. Yet the irony is that the word 'individual' means literally 'undivided', 'indivisible', that which cannot be cut. Mankind is an organic unity however much we kick against the notion, or fight for individual freedom. Mankind yearns to discover what is beyond the boundaries of existence we have created for ourselves. Every woman who seduces a man is enticing something beyond him and within him, where there are no boundaries. She keeps Eros binding and loosening; she keeps Eros shifting, seamlessly sewing, and ultimately eroticizing the body of mankind itself.

Of course there are many people who refuse to be seduced, deny that they can be seduced, simply because they would rather live in the illusion they know than fall into an illusion they don't. Denial of the hunger for seduction is anorexic egotism.

Who are the archetypal seducers? Every woman has the power to seduce, and so does every man. But the quality of enticement has been labelled 'feminine' and there seems to be an assumption that women carry the weight of seducer.

The origins of the word 'relate' means 'to bear' and 'to carry'. We must carry each other and bear each other as the seducer and the seduced. There must be one who captivates and one who is captured; relationships would not happen otherwise. Woman have been perceived as bearing it all, for 'masculinity has come to mean the rejection of the erotic as effeminate, leading to a compensatory coarsening of Eros or to substituting sexuality for it'.[4] Eros is woman's consort, and like Aphrodite she sends him out as her ambassador to court and lure those to her. Woman's image has been perceived, like Aphrodite's, as the deathless goddess of seductive passion.

Seductive Men

Seducers are usually women, simply because a woman accommodates more readily the symbol of the feminine principle. Anything which is deemed contrary, left-handed, moonlit or seductive is considered feminine. The Don Juans, Valentinos and Rhett Butlers of the modern world are usually perceived as un-masculine, effeminate time-wasters to society, or as chauvinistic arrogant womanizers.

Enticement of man by woman leads to honouring life through the sexual union in which the experience of seduction usually results. Enticement of woman by man is just as viable, yet the organic living system of mankind seems on the whole to favour man discovering his vulnerability through woman, and for woman to find her power through man.

In a sense, when a woman seduces a man she is luring the gods out of their sleep, our deeper realms are disturbed, nudged, awoken and arise. It is the dark demons or ugly monsters of our unconscious who bewitch and beguile us. Likewise, woman's image is one to which men are drawn to perpetually. We are all fascinated by our own depths, the feminine growls up through our inner world, masculine looks down and grabs what it can from within. Opportunist seizes harbourer.

When a man meets a woman he may be either seduced into denying his own earthly darkness and imagining it all to be hers, or she may bring torrid Circe and Calypso alive in him. Sexuality then becomes the manifestation of the seduction. This seduction, this romancing of the senses has to become tangible, it must bring into being something vital and living. It has to erotically vibrate with the universe or it will not survive. This energy that has been evoked in our awareness has to take on form, just as our spiritual images seek expression through the world of the tangible in religion and its icons or altars. The pleasure of the seduction is in the senses that are evoked and awakened in the relationship. Earth-bound as we are, sexually we can at least find moments that seem like the ineffable through the instinctive senses of our bodies. But if a man had to carry an eerie image of the 'eternal sensualist' it was inevitable that it was the Sirens who haunted his nights, while woman still haunted his day.

The Sirens

A Siren is 'she who binds with a chord'. The chord is both the haunting song of the sea maidens, and the chord of a resonance between two people; the most binding and magical moment of falling in love. Maya places Siren in every woman, simply because woman is the prime visual icon of the feminine principle. Femaleness is earthly fecundity, inviting and nourishing. The feminine principle when expressed positively is receptive, passive, embracing, when expressed negatively it is duplicitous, deceptive, promiscuous or emotionally demanding. Put the three together and you have positive and negative yin-sense embodied through woman's female nature.

Three aspects become the three sirens, originally daughters of the river-god Achelous and the Muse, Calliope. With the most sensual and mysteriously rich singing voices the Sirens had inherited from their mother, they became handmaidens to Persephone. When Hades stole her away to the underworld the Sirens begged the gods to give them wings so that they might search the world for her. They lived on a remote island at the entrance to the underworld and every time a ship passed by they would sing a hypnotic song that no mortal could resist, in the hope that Persephone would hear them. When the Sirens realized the ship did not carry their beloved goddess, they would drown the sailors or tear them apart with ferocious claws, sending men's souls unburied to the underworld.

Enchantment is both compelling and destructive. The Siren perceived in every woman constellates a spellbinding, drowning-like image, in many men. He may be drawn to her watery world. The power of the sea and her ocean of rhythms, both dangerous and gentle, are potent symbols of his own longing to return to the safe waters of mother's womb or to the unknown, less safe mystery of the cosmic tide. Deliverance is in the enchantress's power, he both yearns for her and desires her, for she may take him and ravish him there.

This profound inner need in human nature is the same for both men and women. Disenchantment means we can be enchanted, disillusionment means we can be seduced. Mystery *is* woman, has been a force-fed tube in our collective stomach. Women are perceived as sensual and seductive, because they embody not only femaleness but all the images of the feminine. It seems seduction has been allocated a 'warning' label. To be the seducer, however, is also to be seduced by one's imagination and the energy that glides through us all. This quality or energy experience is rarely a deliberation, it is not a logical step derived from a process of reasoning. However, depending on our personal psychology, some people may be more responsive, reactive or defensive about seducing or being seduced.

The Seduction of Certainty

Modern-day sirens walk the streets often unaware of their haunting song. Some walk the cat-walk, some dance before the screen, some merely take lifts to the third floor and find they have seduced their way to the top. This is where seduction becomes induction. Conscious beguilement is an alluring sense of power. Media pens scrawl out adverts about the joy of being seduced. Seduction is a commodity, so is sensuality. Advertisers seduce you with their words, they are certain your life will improve if you buy their goods. They are certain your body needs improvement, they are certain you can restore your love life, they are certainly seductive. Certainty is, in itself, seductive. Certain derives from an ancient Indo-European word meaning to cut, to separate. We are seduced into separateness. Cut yourself off from the one-ness of the universe and you will be redeemed, be separate, improve, harmonize. If you buy into this dream you'll achieve, you'll be taller, thinner, live longer, you'll separate yourself from the dreaded nightmares of mortality. You will be seduced away from the limitations of the body, separated and certainly seduced.

The Seductive Message

The attraction of another to another seems threefold. First, a sensory instinctive message relays an image to entice or be enticed; second the mysterious unknown recognizes itself in another; and third, our own images are reflected, baffling and intentionally self-awakening. The triad does not work chronologically, neither of the three works more potently than the other, they work in unison, simultaneously and spontaneously. There is no hierarchical structure, for these energies are the unity of attraction, not of value. But every instant is synchronous with every other instant. There is timelessness in the quality, as the universe in its entirety embraces us.

The seductive message is received with different reactions depending on who you are, your own unique expression of life on this earth, or what psychological tricks and pretences you express or rise to. For example, your sense of separateness is different from my sense of aloneness. The call of the siren will be a different tune for every individual, coloured by each individual's own sense-perception. But it is still a haunting.

Ancestral and archaic memories, the inherited layers of beautiful dreams or dark nightmares and the impressions of generations, and personal thoughts about seduction and the feminine will shape our call to seduction as either threatening, or inviting. The seductive immersion tank holds pleasure-waves in suspension. There we may bathe in the realm of mystery, soak in the desire, wriggle our toes in the enchantment. Whether we are the seduced or the seducer, the exchange of energy is unpredictable, wild, an expression of awakening vitality. It is an unbinding, an 'outrance', a lure set by the soul of the body of mankind to keep us awake to our sleep. It is fundamental human nature to take pleasure in seduction, to be a part of the dance. To be lured out of yourself, to be taken away from the grimace of mortality, the routine and the monotony of life. This is enchantment. But many deny they seduce or can be seduced. Self-denial is the seductive art of self-willed persuasion.

A woman is labelled 'seductress' simply because she may symbolize to a man the sensual art which he disowns in himself. The female body is undoubtedly built to make love, to possess, ravish, coax and inflict both arousal in a man's body and to receive an immediate response. Women are seductive, enchanting and beguiling because they have to be, and also because they are *imagined* to be. Sensual seduction is a powerful invocation to life, as long as it is not done with the will to power or exploitation.

Any woman or man who has made it to the top of their profession, art or skill has probably had to seduce someone or something along the way, or been seduced themselves by an idea, notion or sense of purpose. The 'sensual', like love, needs to incarnate in the topsy-turvy ambivalence of human relationships. These archetypal qualities are evoked through our psychological needs and complexes. My response to a seductive opportunity to make more money, for example, may not trigger the same sense as for you. The 'sense' that gets triggered may be my sense of freedom, or your sense of security. It is only if we consciously back-off from seduction and our sensual nature, or manipulate 'it' without engaging fully in the process, that it becomes dangerous, painful or destructive. Used with wisdom, the art of seduction generates life, and gives us the chance to reflect upon our own natures. Used without awareness, the power of seduction can enchant us into disappointment, resentment and envy.

Cleopatra

A compelling enchantress of two powerful men, Cleopatra, has ironically also seduced mankind's imagination for hundreds of years.

Gautier described her as 'the most womanly woman and the most queenly queen, a person to be wondered at, to whom the poets have been able to add nothing, and whom dreamers find always at the end of their dreams'.[5] Following in her footsteps is rare, yet the penthouse suite lures many. The turrets of wealth and ambition seduce women and men into seducing other women and men to achieve, to succeed, to acquire power.

Cleopatra's beauty and charisma came to represent the dark, manipulative power of human nature. She has been an easy canvas for the projections of writers and artists for hundreds of years, their own underlying motivations, found her an easy scapegoat for sexual confusion, neurosis and the changing perception of who and what is woman and why she aggravates the imagination. The most interesting thing about Cleopatra is that whoever describes her is colouring her with their own lop-sided perception, this writer included. She has been seen as *femme fatale*, seductress, killer, and lover. Embodying all the sensual wiles of the East, the exotic and the unknown, Cleopatra has been one of the West's most provocative 'thought-*fatales*', not simply a *femme fatale*.

Plutarch described her as 'bewitching'. Pushkin, after a fourth-century Latin narrative by Aurelius Victor, wrote how she prostituted herself. One single night of sexual passion with her would be paid for with the man's life. Shakespeare on the other hand, one of the most sensually aware men ever, wrote:

"The barge she sat in, like a burnished throne,
Burned on the water; the poop was beaten gold,
Purple the sails, and so perfumed, that
The winds were love-sick with them . . .'

If we are to have an early archetype of the seductive and yet deadly power perceived as belonging solely to woman then it must be

Cleopatra. Yet it is ordinary women who pay the price for the spellbinding images locked in our psyche. Those women who must express seductive power because of who they are as individuals, are always easy targets for our own complexities and shadows. Jung wrote, 'Every man carries within him the eternal image of woman . . . This image is fundamentally unconscious . . . engraved in the living organic system of the man. A deposit, as it were, of all the impressions ever made by woman'.[6] The seductive woman in any office, home or night-club thus becomes the sensual prowler.

Woman too has an eternal image of woman, as she does of man and man of man. Shaped by collective notions these images mould into personal formations depending on each person's own innate, diverse and multiformed personas.

Sexual Seduction

It is our inherited belief system that seduction leads inevitably to sensual-sexual pleasure. It has been asked, for what other reason would any woman, or man prey upon the other? This became the focus of much of the *Angst* of the 1900s when, as now, men's and women's attitudes to one another were radically transforming. At the end of the nineteenth century predatory females tumbled out of art and literature with an intellectual voice, as men battled with the new women emerging from the machinations of a progressive civilization. Power, extraordinary scientific achievement and abstract growth had reached the midheaven, the point from which is must shift towards the nadir, the downward spiral and cycle of growth and decay. The siren was apparently climbing out of her underwater grottos to drag man down into immorality and primal passion. Sexually insatiable woman was likely to be lurking in every drawing-room in the middle-class West. Were these sober, intellectual and polite rooms of the decadent powers harbouring the siren's song behind the governess's pince-nez, or was it to be seen at the piano lesson behind the alluring innocent eyes of the dutiful daughter?

Artists such as Burne-Jones, Draper, Auburtin and Moreau spelled out the physical allure of the sexually dangerous woman. To be seduced from man's rigorous spiritual strength and morality was to be woman's responsibility, not man's. The fantasy of being seduced by a siren, man helpless in her lair, vulnerable and weakened by her insatiable passion and lust, was one pleasure that men could indulge in and still imagine he was the innocent victim. His own sexual appetite and weakness was thrown literally upon woman's shoulders. Now he was defiantly the seduced, never the seducer.

Passion was firmly attributed to the world of woman, her predatory nature both alluring yet alarming. The cold waters of the sea siren were as binding and as compelling as the waters of her womb. Not dissimilar to the new independent women, the siren was likened to the viragos whom men both feared yet sexually desired. Many men still carry this image of the voracious female, there to lead him into the hollows and caverns of a watery death, his worst fears and his greatest fantasies vividly imagined in every woman he meets. We need to look at how she might be perceived now.

Lisa, all alone, dresses to kill. She has imagined him already, but the bath was not deep enough for her darkest devices. Pulling out the plug she gurgles in tune with the gurgling water. Black voile swirls around her breasts. She has perfected her image in the mirror. It is time to seduce.

'Take me sharply, take me quickly,' she whispers to the mirror.

She, like many women, has a successful career. Single, independent, culturally attractive and in society's eyes nocturnally gorgeous. No feminist, Lisa lures men to her. They seem to want to know her mystery. She dresses and adorns herself, perfumes her body, indulges in flirtation and body language and doesn't pretend to herself that every man is phallic-brained. She believes in equality, both professionally and in intimacy and the fact that men are as vulnerable as she is strong. She respects men as much as the men around her respect her. This night though, she will be different, because there is one man she wants to seduce. Why? Oh, she does not say, not yet anyway. It may be just sex she wants, bodily pleasure, or it may be that she wants to dance with danger.

She spots him across the crowded room, an empty glass held against his cheek as he chatters and flashes his sex-hungry eyes at a younger vivacious woman. Lisa knows he knows. Lisa has practised seduction before. He glides across to the bar, orders another drink. At the bar she brushes past him, and her expensive perfume, unfailingly recognizable, drifts into his sensory range. He turns his head towards the vivacious girl, pretending, oh so carefully pretending, he does not know Lisa is there to have him. Is it he who is doing the seducing now? Is he Don Juaning her, a brute in an Armani suit, effusively poised and efficiently enchanting her?

So she orders her own champagne, takes a stool and flashes her eyes at the barman, bespectacled, tubby, sweating. Sex-hungry eyes, a curse of birth on his right cheek, not her style, no fierce longings here. She laughs, works her smile to be sexual, teasing, alarmingly pagan. This time her provocation is paying. Aphrodite has been propitiated, she flashes her girdle of desire. Not to the barman, who can blame her, but Don Juan begins to notice, he is mesmerized, enchanted, drawn to looking at her again and again. He will touch her soon, she knows it.

But what if we change the script. What if Lisa is plain in society's fashion-conscious eyes? What if she doesn't dress to kill, doesn't demand men notice her, is ambivalent about sex and prefers the idea of a knight in shining armour seducing her out of lethargy and into romance?

Lisa, all alone, bathes all alone. Crash-diet number twenty starts as she stares at her reflection. She wants to be loved for who she is, not what she looks like, isn't that real equality? She powders her nose, slips into a tighter skirt than she would normally wear, but leaves no doubt about her statement that she is un-made up, natural, tousled and bra-less.

There is only one stool at the bar; she sits uncomfortably alone among the many she knows. The barman makes innuendos, leers at every pretty girl and sweats into the punchbowl. Lisa yawns, why did she come? For company, for the knight to find her? Then she sees him, beautifully dressed, refined, gracious. Seducing women would come easy to him. She stirs her drink with her finger, brushes her hand through her hennaed hair and wonders how other women can be so overtly seductive. She crosses her legs, sways a little in time to the music, bends to pick up a dropped coin and he notices her. Something sensual, something mysterious in Everywoman is more potent a power than cosmetic fallacy. There is something trembling, yielding, earthly and yet terrifying about her. She does not know it yet, but she has seduced him. She does not know it yet, but the moment will come when he bumps into her arm at the bar and apologizes, smiles, grins his sexual welcome. A passive seduction becomes an alchemical silent liaison.

If we are honest, no one is exactly like any of these characters. They are merely personifications of our imagination, but for all that they often mysteriously manifest thenselves as real live people. Because of our own suppositions we may assume that the first Lisa is bound to seduce any man she desires, and the second Lisa is less likely. Yet every woman, like every man, is different, and one's own expression of the 'sensual' seductive language will manifest in a different way. Like Cleopatra, who is of fire and air, giving her other elemental qualities away to 'baser life', whatever a woman's own psychological make-up and needs, she may not resemble the image of Lisa number one, nor even number two. Her own perception of whether she is a seductress or is seduced depends on her own interaction with relationship itself. If we become defensive about our sensual awareness we may find difficulty in expressing it.

Why Do Men Need Seducing?

The soul of the world would not move so mysteriously if there was no seduction. The mysterious unknowable sets lures and traps to keep us reaching out for the other. It is not so much that women seduce men, but that seduction, the energy which forces two people to interact, permeates the relationship. Through the potency of a man's primal sexual urge, by enticing him to her biologically, woman unconsciously ensures those energies or forces which must be eroticized into life are given a convenient entry point. Woman is biologically created as a harbourer, a receptive invitation. Whether she lures passively or dynamically, consciously or unconsciously, her sense of that seduction may be that it is 'herself' or 'himself' as seducer. But more mysteriously it is seduction itself which animates the connection.

Men's own 'yin sense' is often a complete mystery to them, and they project this unknown and unknowable quality onto woman: enticement of the mystery within men, the life-force that needs awakening means they are fated to move, to be in relationship and to keep the world in motion. Seduction has the power to eroticize and entice that which is beyond mankind, through the interplay of a man's sexual potency and a woman's sensual love.

A man unconsciously asks a woman to carry his own sensual nature every time he encounters her, unless of course he is very aware of his sensual expression. He secretly calls out for her to be the sensual and for her to entice him. He, depending on his own particular sensual responses, will seek a version of sensuality that will awaken love desire, body desire and soul desire within him. Alternatively every time she encounters him she unknowingly asks for him to be like the sensual image she has imagined, and negotiates through her seduction or enchantment of him a way to incarnate the mystery, which through sexual pleasure may manifest as a child of that union. But seduction carries heavy fines for those who exploit its power as we shall see.

I lured him out of his senses, I lured him into his senses;
I loved him to his bed.

Notes

1. Hillman, p.282
2. Wilber, p.40
3. Rilke, cit. Baring and Cashford, p.669
4. Hillman, p.671
5. Gautier, p.8.
6. Jung, p.50

PART 6

Pleasure Hunt

CHAPTER FIFTEEN

Femmes Fatales

'Glamour can delude
Power can seduce
Mystique can challenge
Wickedness can tempt'

The sensual is provocative, and its most tantalizing expression has manifested itself in the *femme fatale*. The *femme fatale* is the woman who displays and is haunted by the darker side of 'feminine' images, intrinsically woven into us all, both men and women. 'The archetype of the *femme fatale* began in prehistory and will live forever.'[1] She is an enticing symbol of mankind's greatest fear, our unknown fate, and she is usually a 'real' woman. Personifying certain shadowy aspects of goddesses such as Kali, Circe, Hecate and Medusa, her character is of a woman-beast, beguiling, sensually alert, and of the night. The *femme fatale* embodies the equivocation of sexuality as both seductive and destructive.

What is it about the *femme fatale* image that fascinates, yet is potentially destructive? Who is she, and is she in every woman? Are her sensual wiles and sexual ambivalence of value, or have we merely twisted the woman of passionate darkness into an easy scapegoat for the denial of nature's beast within us all?

What Is She?

The fascinating and beguiling attraction of the *femme fatale* image holds us spellbound. Many wish to demystify her, to grind her up, to dissect and tunnel into her paralysing presence. But to do so is to deny the ambiguity of nature's mystery. She is the personification of mystique rather than the personification of mistake, for she embodies the primordial dread of mortality. A screeching Lilith, a seering Medusian terror, a strangling 'Throttler', a Salome and a Judith, all are perceived as possible characters embodied in the *femme fatale*. Unknowingly she has carried these raging qualities which lie as dormant impulses and experiences within both men and women, laced with the power of female sexuality.

The demonic femme fatale is a symbol of erotic fate, both to mankind and herself. Her sensual world engulfs ordinary sight, sound, taste, touch, smell, and expresses the deepest hunger for love, for hate and for feeling. She is instinctual, famished and thirsting for sex or transformation. She may also be hungry for love.

She may become manifest as the Lamia, fatal to her children; she may be the Throttler, fatal to those who are not wise; she may be Clytemnestra, fatal to those who do not love her enough; or Marilyn Monroe, fatal to those who ignore her vulnerability. Or more recently Diana, fatal to those who deny their own weaknesses behind the trappings of power. But the irony is that the *femme fatale* is fatal mostly to herself.

Instinctive Fate

The instinctual side of our nature is often alien to us, a layer we experience sometimes with little awareness down beneath the hallways of politeness, bellowing in the dungeons beneath the boudoirs of rationality, logic and social expectations. Instinct groans upwardly into society, pushes her head above water like a mermaid, a siren, a Kraken beast or Grendel's mother. Our psychological dynamics work quietly in the pit-stop, the intentional force of nature, the life-force, keeps us moving erotically, tuning our forks to the beast within. Beneath the mechanical whirr of society and acquired culture we are all instinctive animals, and it is through our sexual hunger, our need for bodily contact and sensual love, that the most provocative *femme fatale* in every woman may push her claws up out of the subterranean hollows. Inherited curses, skeletons in closets, family fables and collective legacies coagulate in these murky depths of seductive longings. In the intellectual prism of rationality, a woman's sexual power feeds and waters *femme fatale*'s dangerous allure. The dark feminine finds a temporary nesting place in those who side with Cleopatra's asp.

Who Is She?

The Jane Mansfields and Bardots of this world have been sentenced to a mortal life as immortal sex goddesses. Bardot at least escaped the firing squad, and no longer attends, or pretends to be anything other than a woman. Mansfield's reign was short lived and forgettable. But the sex idol is not a *femme fatale*, she is a constructed embellishment of mankind, a twirl on the end of humanity's ice-cream cone to satisfy the desire instinct, a contrivance to tap the power of female sexuality as a commodity. To be a mortal goddess is an anathema. Anathema derives from a Greek word meaning, literally, 'that which is set up'. The media sets up the woman to be a goddess, but real goddesses are immortal, they are not human. To achieve immunity from mortality you must either die young, or relinquish responsibility for human love, choices and feeling.

Femmes fatales are living females, not mortal goddesses. They live and breathe mystery, they are emotional, auto-erotic, deliberately seductive and disinterested in being immortalized. As movie stars, celebrities or powers behind the throne, they may well live out their inner fears secluded under the masquerade of pretence. But they die as well as live the art. These restless spirits are deeply passionate about life, love and meaning. They bear loneliness in quixotic silence, and passion and power become compatible expressions of a deadly integrity. *Femmes fatales* are humans who take responsibility for their choices and actions, which is why being a femme fatale is fatal in itself. They suffer, for they are both sexually powerful and emotionally vulnerable, although they may never admit to either.

There seem to be two types of *femme fatale*: sexually devouring or ruthlessly discriminating. The latter operates alone, frozen in a time and a place where there is no past nor future. It is not so much her beauty that incites, but her almost frigid response that beckons. She is nocturnal, cool, as icy as the winter new moon. Her presence is static, anticipatory. Immobile corrosiveness in a silent gaze. Her looks kill, her eyelids glitter with desire.

Christine Keeler was a recent *femme fatale* involved in political intrigue and scandal. She began an affair with John Profumo, the British Secretary of State for War, while already involved with a Soviet agent. Her affair with the osteopath Stephen Ward, renowned for introducing beautiful young girls to wealthy patients and influential men, eventually led to his suicide. Profumo was forced to resign.

Then we have the other *femme fatale*, she of the vampire, of the full moon. Like Lilith, like the Lamia, she seeks vengeance or is hungry for blood love. Sexually ruthless, she is ready to swoop on her prey with precision and timing. She may have been wronged, jealous, doomed to pay a high price, she may have chosen power as her weapon or her defence, but she must live out herself. These are women like the sixteenth-century Reine Margot, ambitious, beautiful and powerful, who had much influence over the royal family of France. Notorious for her many love affairs, after her marriage to Henry IV of France was dissolved she returned to Paris and built herself a love palace. There at the age of fifty-five she favoured a youth of eighteen to her current lover, the Comte de Vermond. In a duel, the Comte killed her young lover, and in her revenge, Marguerite de Valois cut off his head.

In whatever form she incarnates, this archetypal woman appears to paralyze or castrate the intellect or physical potency of a man. But always she seems destined to render him insensible with her own senses, as a wolf spider paralyses its prey: a man's most instinctual fear.

Why Men?

Why is is that men are so compelled to possess her yet so fearful of her power? It seems that the *femme fatale* personifies all aspects of the 'feminine' cloistered and accommodated in the female body. If the female body is the most fundamental symbol of the feminine, or yin-sense, then this fatal sensual energy becomes intensified, exaggerated and distorted by her very presence as dark, light and magnificent. She holds out to men a set of traffic lights which, depending on his own psychology, will either say, stop for a while, hesitate if you dare, or go on. He will be repelled or attracted, drawn to her, or run from her. He may project onto her whatever he disowns, finds uncomfortable, or secretly admires in himself, usually in the form of his own feelings, envy, vengeance, passive aggression, jealousy or fear.

Collectively, mankind has gilded the *femme fatale* with mystery, power and fate. This golden triad is nature's divine secret, and the twist in the tale is that the woman who lives out her dark angel is, like the seductresses of the last chapter, enticing something beyond mankind. Women's femaleness magnifies and embroiders her sense of femininity, just as man's maleness inflates his sense of masculinity. Woman's female nature induces the sensual into being, as her body induces a child to be born. The feminine is about more than just passive preservation and cooking buns, it is about the itch that can never be scratched, it is about reminding humanity of its darkest hour, and about reminding men how vulnerable they are behind their physical strength and cerebral *gravitas*. The *femme fatale* is the burning barge of Everyman's past.

'Fatal Women'

Aphrodite's dual nature certainly embodies the archetypal 'goddess fatale', with her vengeful, scheming, dangerous liaisons with mortals, demi-gods, nymphs and gods alike. But it is those who are not deities who maintain the recurrence of fatal attraction in history.

The fantastic Throttler, or She who Strangles, was also better known as the Sphinx of Thebes. Although not a complete woman, she is one of the earliest *femmes fatales* in Greek mythology.

The Throttler

Above the gates to the city of Thebes roamed the Sphinx, not the Egyptian Sphinx of the tombs, but a creature with woman's head and lion's body, adept at terrorising innocent passers-by and those who wished to enter the city. Whether she was sent by the gods to ensure Oedipus's destiny or not, she was ultimately responsible for Oedipus marrying his mother. The Sphinx was a deadly *femme fatale*. She posed a riddle to each man or woman who entered the city. Wrong answers meant strangulation. No one answered correctly until Oedipus came along. The Thebans' reward for ridding the city of the Throttler was the hand in marriage of the recently widowed Queen, none other than his own mother (unknown to Oedipus). The riddling power of this static, paralysing, half-beast half-woman who lured men to their death is no different from the sexual equivocation behind other *femmes fatales*. And like most fatal encounters, this charismatic and ruthless creature came to her own desperate end when Oedipus answered correctly. In her despair, she threw herself off the top of the city walls and crashed on the rocks below. This self-destructive twist is the lot of many a *femme fatale*, especially human ones.

The modern Sphinx too may lie basking in her basque. She may emerge in ambitious, provocative, self-confident women, feminists or not, or those who have discarded their passive femaleness for the sake of potency, progress or power. Women are demanding of themselves the equality that is their right, but they may be clutching the phallus too firmly at the expense of the vulva.

The Throttler may be seen in boardrooms, the media, the educational system, in politics and even in religion. She is fatal because she uses intellect and cerebral potency as vivid and slick as any man. She weaves her riddles through the outer chambers rather than the bedchamber, but she may use her sexuality when necessity beckons her. The Throttler plays her word games, poses questions, never answers. She is immaculate, well-groomed, a wild cat with culture's claws. But is she a flagrantly sexual woman beneath the pin-stripe suit, or does she truly deny her femaleness? It is not sexual sense that she uses to trap her victims, it is the sense of knowing, of *making sense*. Knowledge is powerful, and there are many *femmes fatales* who radiate a sensual eloquence and articulateness, a quality of sensory insight and wit rather than physical grit. If we do not make conscious the underlying motivation for our need for power, whether we are men or women, the sense of wisdom can be exploited just as easily as the sense of sexuality.

Phryne

Another stereotype *femme fatale* is the one who knows her sexuality is a power to be honoured. Phryne was a Greek courtesan in Athens around the fourth century BC. Her dark complexion and looks gave her the name, Phryne or toad. Mistress of the great sculptor Praxiteles, she was also extremely wealthy and offered to pay for the rebuilding of the walls of Thebes after they had been destroyed. She had many lovers, was accused of defiling sacred religious worship, then brought to trial to be hanged. Another of her lovers, Hyperides, was her lawyer, but even he could not argue her defence. Whether it was Phryne herself, or Hyperides who suggested it, she used the power of her sexuality to persuade the judge otherwise. Unclasping her chiton and drawing back her under-robes, she revealed her naked body to the judge, was immediately acquitted, then carried in triumph to the Temple of Aphrodite. There is one account which suggests that her lover Praxiteles, driven by jealousy, passion and pride, could not bear her immodest reputation and finally slit her throat at the foot of the great statue of Aphrodite, the very one Phryne had modelled for him. Women who live dangerously often die dangerously. Phryne is a *femme fatale* of the sexual and erotic senses. What draws men to lust, debase and violate women is usually of instinctual, archaic origin; it is the same love potion as sexual desire, but offered in a different Grail.

Mata Hari

Although much biographical work has attempted to redefine her actual influence and fated enchantment of her lovers, Mata Hari still conveys an enigma, a woman who may or may not have been a double, triple or quadruple agent. We respond to the name as we respond to the word 'sensual': we cannot separate Mata Hari from 'dangerous woman', as we cannot separate 'sensual' from sexual woman.

Mata Hari, or Eye of the Dawn, was a professional seducer of men. Like Eos, the goddess of the dawn in Greek mythology, who was cursed with lust for young men, Mata Hari appears to have a long list of lovers as her conquests. She joined an exotic and erotic dance company in Paris, symbolic of sacred worship and possibly similar to the sacred sexual Dance of the Seven Veils. Mata Hari loved and desired men across Europe, mostly high-ranking military chiefs, politicians or government officials. In 1907 she allegedly joined the German Secret Service. In World War I she became infamous as a double agent spying for the French in Belgium, but the French were suspicious of her liaisons with top German officers, and in 1917 she was brought to trial and executed as a spy. There is still much controversy surrounding her life and her espionage. Another *femme fatale* destined to meet a very tragic fate herself.

Women who purposely court mystery and physical danger, such as Aphra Bhen, another seventeenth-century spy, Charlotte Corday, the French revolutionary and Marat's assassin, Mata Hari, Lucretia Borgia or Eva Braun, are not only expressing their dark feminine energy and potent life-force or spirit, they are engaged in the intricacies and tricks of life, of the movement of the illusions of life, of seeing below the surface. *Femmes fatales* are not just courtesans and men-seducers, they are usually those who have a breath-taking passion for life, a passion for passion and the dark hallways of danger. Like those involved in the arts, creativity, psychology, history or anyone fascinated by human nature, these are women who live the art of the web of life as they weave it. This is the *femme fatale* of the underworld senses, risking one's life for passion and fate itself.

Dangerous Zone

The *femme fatale* is dangerous, for she has become the most potent reminder to mankind of mortality. But what is danger? Danger derives from an old French word, *dongier*, meaning absolute power of the feudal lord, which in itself originates from the Latin word *dominium*, ownership. Dangerous means resistance to the absolute power, and is disastrous to man. It is easy to glimpse how the *femme fatale* has become a malevolent spectre in the masculine dream. She seems to threaten the absolute power of traditional patriarchal dominance. So *femme fatale* equates with woman's resistance. She appears to, and often does, defy men from her icy world. Her fixity and her Medusa-like petrifying gaze are nagging reminders of mother, or terrifying nightmares of death itself. She is a dangerous zone, but always erotogenic.

Ordinary Fate

In literature, one of the most ordinary but androgynous of erotogenic women was Emma, the heroine of Flaubert's *Madame Bovary*. Emma is doomed to disillusionment, a common fate for those with high ideals and expectations of love found in romantic novels and the advertising romance of today. Emma, married and bored, takes up the call of the mid 19th century woman. As a participant in European patriarchal culture, and floundering in a marriage which seemed to hold none of the expectations she had imagined, she resorts to romantic and scandalous encounters. She even dresses as a young man at soirées and dances. But her affairs become disillusioned. Her desire for something mystical tinged with sensual dooms her to a tragic, suicidal end.

Flaubert, in his many letters and papers, admits to his own inner sexual contradiction, an echo of Emma Bovary's irreconcilable longings. 'There comes a moment when one needs to make oneself suffer, needs to loathe one's flesh, to fling mud in its face, so hideous does it seem. Without my love of form, I would perhaps have been a great mystic.'[2]

The anxiety of the nineteenth-century male, no different from now, bound by society's own expectations of him, was overwhelmed with psychological complexes around his inner feminine image. Mysticism was then and still is labelled a 'feminine' quality. There was still little awareness among middle-class parlours and polite society wives that there was a difference between sexual pleasure and romantic love. Eros was regarded as a boy with a limp penis, and the hierarchical roles of men and women, synthesized as they had been since Christian churchmen found the power of containment could be disguised in the marriage vows by the word love.

Emma is an ordinary woman, a *femme fatale* of polite society. Today she may be the career woman, the mother, the writer, the beauty therapist, the bored housewife, the radical feminist, the teacher, the nurse or even saint. She is many-spectred. But Emma tires quickly of sexual passion, and yearning for yet another affair, begins to prefer her imagined romantic fantasies to the real thing.

'They knew each other too well to experience, in the act of possession, that sense of wonder.' Yet like the romantic fool she is made out to be 'she still clung to it' – the affair – 'from habit, or perhaps, from some perversity of her nature. Each day that passed she pursued him with unrelenting ardour, spoiling what delights she may have found by setting her hopes too high . . . she even longed for some catastrophe, so that it might force them apart'[3]

Sexual pleasure is not enough. Emma needs a more powerful erotic transformation, pain, suffering, death and danger. Flaubert too may have been searching for his mystical 'sense of wonder' – another sense of romantic unworldliness, which cannot be sustained by familiarity.

Emma Bovary exploits her sensuality, and her lover Leon only becomes real to her as, 'a synthetic figure compounded of things remembered, things read and loved in books and the images created by her own insatiable passion. This shadowy creature became at last so real to her, so accessible, that her heart beat faster even though she could not see him clearly . . . These visionary ecstasies of love left her more exhausted than whole nights of dissipation'.[4]

Gradually she comes nearer and nearer to her own ghastly end. Through her tragic dismay at trying to borrow money from her ex-lovers who refuse because they simply don't have it, she realizes that sexual Monopoly is not enough to save her. What she needed was love. Poison is Emma's answer to rejection and her own self-sabotage. Like, Cleopatra and her snake, Emma Bovary looks for a way out to end her agony. Like Mata Hari, Emma lived out a similar self-deception, but even her enchantment is not enough when she reaches the bottom step of the pool. Attempting to seduce Leon into stealing a huge sum of money for her to save her family from bankruptcy 'the expression of her flashing eyes was brazen, devilish; the way in which she kept her lids half-closed lascivious and provocative'.[5] Her other lovers too refuse her pleas. Flaubert's Emma is now a wild woman, nature's whirlwind, her insatiable longing for passion now transmuted into longing for death.

Flaubert has painted Emma Bovary as sexually dangerous, provocative and mysterious, an ambivalence between spiritual devotion and sexual urgency. This says much about Flaubert's own inner dichotomy, and the traditional anxiety around sexuality at the time. But it also displays the notion of many men caught up in the underworld of their own sexual confusions about themselves and about women. Flaubert polarized the fear of fate in himself into Emma, just as many people still assume fate and 'the sensual' as exclusive to woman.

The deadliest twist of all is that the *femme fatale* symbolizes mankind's own fate. Locked in her embrace is the mystery of the mutable through mutual participation. For many men, the insecurity of the ephemeral world brings fear of sexual exile. He believes he is as banished from nature's passion as woman appears to be its protagonist. He is an outsider, doomed to be drawn, shocked, awakened into nature's ambivalent arms whether he choses so or not. His own humanness seems most vividly contrasted by the *femme fatale*. A woman's physical presence can arouse him into his violent spasm of desire which is then drawn from him in the darkness of her body.

The ecstasy of love that Emma Bovary wanted and yearned for was not real, earthly passion. Idealistic love and her cultural confusion around sensual pleasure became tangled beneath the bedsheets, and is still a common dilemma for women. Emma may be an allegory, a pathetic, doomed woman, but there is a deep truth behind her fatality. She is the *femme fatale* in Everywoman and Everyman. Sensual love is physical pleasure; romantic idealism is unworldly, transcendent, numinous. The dilemma of unifying these two aspects of love in our monogamous relationships, as we have seen, is the most fundamental expectation of all. The *femme fatale* has no other choice but to die because it is the most profound statement of herself and her hopes. Like Cleopatra, Virgil's Dido, real-life Mata Haris, Eva Brauns, or the Greek favourites Medea and Clytemnestra, the price women pay for living out the archetype is living on the edge of life, rather than living on the edge of fear.

Masochism

Another view held by most nineteenth-century men, and one often still held today, is that the *femme fatale* is an inflictor not only of desire but also of physical pain. This was one of the first attempts at defining her, a fashioned stereotype which still feeds our minds. The word masochist derives from Leopold von Sacher Masoch, who among other things, wrote towards the end of the nineteenth century about men who were obsessed with women who inflicted pain upon them. Here the dangerous woman was suddenly at her most powerful. She was allowed out from her cage. The salons, tea rooms and middle-class drawing-rooms and parlours were exciting meeting places for those who might harbour a whip, a thong or a kitchen knife under their petticoats. The *femme fatale* became dangerously close to death-watch woman.

Venus in Furs was Sacher Masoch's most perverse offering, his hero Severin a sucker for punishment. Through the enslavement of his body by women, the hero supposedly transcends the baseness of his sexual desire. Severin even admits first to falling in love with a stone statue, the most hardened and experienced of *femmes fatales*. 'Often I visited that cold cruel mistress of mine.'[6] For Severin, 'woman represented the very personification of nature . . . she was cruel, like nature herself'.[7] Severin's 'supersensual stimulation' was man's manipulation of woman into a brutish slave-driver who would redeem him through punishment. Thus making him feel that his sexual desire was being subjugated safely away. She would take responsibility for his passion, and equally if he played the victim to her aggressor, be relinquished of his guilt, as common now among sexual power games as it has always been.

From the erotic flagellation of the medieval inquisitors, to the modern day porn movies, men have unconsciously colluded with the archetypal *femme fatale* to relieve thenselves of sexual anxiety. Gautier's 'Cleopatra', Moreau's painting of Salome, or the decadent fascination for Judith and Delilah, were vivid images of the cruel and barbaric woman that obsessed the intellectual and artistic circles of the *fin de siècle*. The perception of woman had not changed, only the content of her ambiguous power.

In early Victorian England Mrs Berkley became infamous for inventing her 'Berkley Horse', a device to which her male 'clients' were strapped for various spankings, flagellation and guilt-free sexual pleasure. The face and genitalia of the customer were projected through well-located spaces. Whip-lash woman, now apparently able to wreak vengeance on patriarchal conditioning, would flagellate the fated man while another woman, according to the customer's personal taste, would stimulate his sexual parts.[8] The only difference now is the video camera or the web-site.

Homme Fatal

There are very few *hommes fatals* because men are biologically unable to sustain the secret ambiguity of their sensual and sexual intentions. Women have carried the dangerous image of sexual temptress, while men have carried the danger of destructive power, simply because of the fundamental differences in our biology. Men cannot be ambiguous about their sexual arousal, it shows and it urges them to act. Women can be ambiguous about their sexual desire, it is hidden if they so chose.

The few hommes fatals there are, Lord Byron, the fictional Romeos, Heathcliffes, Rhett Butlers, the real Casanovas and Rudolph Valentinos, are different from their female counterparts because man, in his most primitive biological form, must keep on the move. The *homme fatal* is basically nomadic, a herdsman, he must cover ground, change the pasture, remove evidence of his camp-fire. He must cover his tracks, change his scent. The *femme fatale* is resistant, vigilant. But the *homme fatal* has no time to lose, his boundaries widen, he marks his territory then spreads his seed. His primitive need is to generate quickly and everywhere, to hunt and find new egg caves, and to ensure he seeds new hunting grounds wherever possible.

The real *hommes fatals* are those who acknowledge the life-force and the soul. They feel, they are overwhelmed by nature and the gods, they are passionate about love and life, and are aware of the sensual quality which they chose to call 'she'. The romantic visionaries in the arts, passionate men who live life on the edge like the *femme fatale*, are no different from her. Men's repressed sensuality bubbles close to the surface, arising in dangerous liaisons, in art, in literature, in power and in painful relationships. Like invisible gas, the 'sensual' is an energy which creeps unseen sending ripples from its unfathomable depths up through the cracks in the cultural rock of society. Both men and women embody this energy, but because of the dualistic Cartesian perception of Western civilization, most of us still cannot see 'sensual' energy in any other way than with a feminine label. The femme fatale may well use her sensual energy to dazzle and allure; she may well take the risk of acknowledging this all-encompassing force. It is only because culturally we have chosen to identify the senses with one gender, that we have created a boundary around 'maleness' with a sign that says, 'sensuality out to lunch'.

The New Femmes Fatales

Who is our everyday ordinary *femme fatale*? Can she exist if she is so wilful, so likely to be unordinary, volatile, extreme in both her feminine yin-sense and her masculine yang-sense? For she will display both, the urge to activate, and the passive alluring coolness of one who will wait. Is this subtle and contradictory blend of life essence not sleeping within every woman? The new *femmes fatales* are women who suffer just as deeply as the archetypal image. The *femme fatale* will always fascinate, because it is the art of the enchantment of life to be embodied through her. Women may enchant better then men, simply because they are the symbol through which the quality of enchantment is best expressed. Enchantment moves through women on a quest, a mission to vitalize the instincts. The *femme fatale* especially conveys the aura of mystique. In every woman is the nuance of danger, of fatal encounter. Taking man into her, he discovers the depths of his own being through her cultural miasma. He must be filled with desire and romantic imaginings to get him past his own archaic fear of woman's body. Enticement takes refuge in a woman's persona more easily, simply because she is biologically the one who must be found and won. The species must survive through her.

The new *femme fatale* may keep leather and thongs in her wardrobe. She may play the efficient secretary by day and the passionate virago of the porn videos by night. But this is only a constructed *femme fatale,* she who is born of cultural need, not nature's uncontaminated amorality. The media-styled *femme fatale* is playing out a role defined for her by society. The real *femme fatale* is one who constellates the 'dark feminine', the soul and the life-force, the hero's spirit and the whips and spikes of ancient growling. She hears nocturnal frenzy singing in the chimes of death-tolls, not just the ringing of glass rims in cocktail bars.

The true *femme fatale* is not society's or even the tabloid's fantasy. She cannot be anything other than who she is, and she will be fatal to others as well as to herself. She may appear in unusual form, or disguise herself according to cultural needs and expectations. One recent example is the scintillating Pamela Harriman, married many times to men in power from the Churchills to business tycoons like Agnelli of Fiat. She became an aide to Clinton, and finally US Ambassador to France. Diana, Princess or Wales, and Marilyn Monroe may be more fated examples. For all their embellishments by the film or media culture, these two women lived short but bizarre and passionate lives, an expression of who they were. Psychologically, the *femme fatale* is vulnerable and driven by the dark demons not only of her family's curses, her society's myths or her social romances, but by the oldest and most primitive of restless forces – the erotic nature of herself. For every relationship she has, each man or woman she encounters will recognize something of themselves within her, but whether they are aware of this recognition or not is another matter.

An Everyday Woman

Take a modern scenario of an old story. This woman, Ella we shall call her, has a twin sister, beautiful, successful, famous for her charisma. Ella feels unloved, a worthless replica beside the glamourous world of Sis. Yet she marries well, has three children and enjoys a well-off lifestyle with benefits. One day her husband has a serious car accident in which her oldest daughter is killed. She never forgives him, feels he is to blame, and he in turn looks elsewhere for love and finds it easily in another woman's arms. Now Ella seeks only vengeance. She is near the edge of life: passion, ecstasy and deliverance could never be compensation for the loss of a child. Then she meets Henry.

'You are the most vibrant woman I've ever met. My god, the passion just courses through your body! Ella, what is it inside you that is so wanting?' So Henry moves in with her.

Ella's struggle doesn't end. She feels fated by her encounter, as if it has deeper meaning. Her son James hates the intruder in their cosy life. James sees his real father as saint, his charismatic aunt a much more appropriate mother figure than his own. Mother is powerful, looks dangerous, destructive, torn by her loss and patched up by sexual attention.

Then one day Ella's husband returns with his younger woman. He takes James on holiday and offers to let him live with them in London, far away from the senseless wastelands of an over-sexed suburbia. Ella seeks more revenge and Henry agrees to do something wicked for her. Passion and intense sexual desire lead us into the darkest of waters and blackest of nights. Together they plot, and murder Ella's husband and new girlfriend, perfectly, with no evidence. Except there is still James, the son. Ella fantasizes once, just once, that she could murder her son, but the power of motherhood holds her back from infanticide. James suspects and fears for his own life. His vulnerability is put to the test. Twisted as life is, and destined as the *femme fatale* always is, Ella is drawn towards her own fate. James hesitates as she screams for mercy, as she tears at his clothes, rips his face with her fingernails, then he thrusts the knife in her neck, her apparent monstrous passion and fury finally avenged across the kitchen table, the plate of chips and ketchup disturbingly untouched.

Not all *femmes fatales* are as extreme as this fictional Ella, or her earlier archetypal killer, Clytemnestra. Yet there is something raw, bestial and demanding in the *femme fatale*, not a graced gift from the gods, not the sensual glamour of the next chapter. The femme fatale has a different sense, she smells fear, hatred, power. She knows fear, knows the instincts of fate and mortality. The *femme fatale* may well exude beauty, charisma and grace. She may be powerful seductive and sexually intense, like Cleopatra or Helen, both of whom we have already seen embody the qualities of many feminine images, but the archetypal *femme fatale* in her loneliness is fated to be nature's greatest example of passionate self-will, and awesome vulnerability. She is danger personified, and for that she is fatal mostly to herself.

NOTES

1. Paglia, p.339
2. Flaubert, p.xv
3. Flaubert, p.282
4. Flaubert, pp.282-3
5. Ibid, p.289
6. Sader-Masoch. p.10
7 Sader-Masoch, p.32
8. Tannahill, p.306

CHAPTER SIXTEEN

Glamour

*'For she was beautiful – her beauty made
The bright world dim, and everything beside
Seemed like a fleeting image of a shade'*[1]

Sensual Glamour

The fundamental essence of sensual sexuality is to take pleasure through physical love, but society's changing prescriptions about love warps and distorts our perception of the sensual accordingly. Whatever the current psychological expression of an era or generation about who and what is sensual or sexual or both, some people seem to embody a mysterious quality that is an evocation of all the senses, in a word which we call glamour. But this sense of glamour has become an easy target for corruption, exploitation and the subsequent devaluing of its meaning. Similarly the glamorous person can be designed by culture and/or exploited by the prevailing vogue. Whatever the underlying cultural motivation for what is and isn't glamorous at any given time, the quality of the truly glamorous woman or man remains sealed in an alembic of self-defiance.

The Meaning of Glamour

But what does glamour actually mean? Popularized and generated by Sir Walter Scott in the eighteenth century, the word glamour is a corruption of the word 'grammar', and its derivatives, 'gramarye', or 'grimoire'. The base meaning of grammar is 'that which is written', and the *gramarye* a magic book of spells, or obscure esoteric writings, the Latin grammar of which was difficult to understand by anyone other than the educated.

Glamour is the mysterious aura that surrounds a person. Some people believe or are fooled into believing we can be glamorized by the media. It is certainly interesting that it has become a commodity many would like to own. Make yourself glamorous by becoming a movie star, make yourself glamorous by using the right make-up, make yourself glamorous by the definitions of fashion or posing as a rock star. Many of us want desperately to be perceived as glamorous because the connotations we impose upon glamour are that it will bring us wealth, success and fulfilment. If people want glamour, then they'll buy it. But what are we buying? Not mystery itself, for mystery cannot be bought, but certainly an illusion created by the magi for our delight and possibly their gain. We buy into this kind of glamorizing because we want to be enchanted away from our disenchantment. The end of the astronomical Age of Pisces has been one of disillusionment, and a search for a meaning in life, one where the materialism so apparently meaningful to us all is in itself only an illusion, which many of us are beginning to recognize as 'not enough'. Being enchanted by man-made glamour makes sense if we are aware enough to acknowledge that Hollywood stars and supermodels are there to promote civilization's illusions, not nature's ambiguity. This illusion, paradoxically, is also an outer expression of a sense of profundity. We are in need of enchantment so we enchant ourselves. The grimoires of our mind, turned into real flesh and blood, are a manifestation of society's needs, which is as it must be, as long as we don't suppose the creations of elitist choice are the real mystery and the only truth. Glamorous celebrities and thin women are an expression of an arcane magical wand twirled between the fingers of financial magicians. The sleight of hand trick is, however, enchantment's ruse, not mankind's.

Charisma

True glamour incarnates as a sensual quality. It is the irrevocable sense of intense magic, of the person who embodies an iridescent invocation of all that has been and will be. You are either glamorous, or not, simply because glamour is charisma, and charisma is a gift given by the gods. Glamour is Aphrodite's blusher dipped into Persephone's beauty box. When a charismatic or glamorous person walks into a room you are awakened to magic, to the girdles of desire, to the musky erotic sense of deeper living. You may be drawn to her or him sexually. Magnetized, bounded and filled with desire, you may despise, hate and envy that person, simply because your own sense of self seems worthless, split, unresolved and too complicated to ever exude charisma. Your psychological complexes say a lot about your perception of who and what is glamorous, and whether you believe you are a charismatic tiger or a mild goldfish.

Charisma is derived from the Greek word *charis* meaning grace. In Greek mythology grace is the attribute given by the gods if you are favoured. Charis was Aphrodite's attendant, the personification of beauty's grace. Charisma has an earlier root, from an ancient Indian word meaning desire. The glamorous person embodies desire. It is a divine sense, an inner holiness, a primitive sacred force, which means walking your own path. Charismatic people are natural heretics. Whether they are aware of the myths and romancing of society or not, they make choices for themselves, and convey a sense of completeness. The glamorous and charismatic person may also seem androgynous. *She* may appear overtly female, she may use her femaleness and her femininity to lure, to attract, to desire, but her masculine qualities will be as authentic. *He* may appear outrageous, potent, arrogant and masculine, but he will acknowledge his feminine image and values, or channel them through art, nature, or the intensity of passion. This bisexual ambiguity is a terrifying attraction to those who believe we are merely man or woman, face-lift or chin-sag, dead or alive. Relationship is the art of living. We perceive the difference between one thing and another, we recognize the difference and give it value or labels, or infuse those things we like and those we don't like with different qualities. Similarly, we can only see ourselves through comparison of what we think we are, with what we think we are not. We create our own subjectivity and then project onto others those very images, shadows and unknowables that we cannot see in ourselves. In this way glamour then becomes someone else's privilege. It never seems to be ours.

Nature and culture are always in dialogue. In history, in the breathing universe they fight or argue, they sometimes tell stories, they sometimes debate, communicate, chatter and whisper who dares. In the charismatic person, nature and culture are inseparable. The body may display femaleness or maleness, but the inner being is alchemical gold, the senses vibrant, mind and soul autonomous and centred. There is a sense of a culmination of grace and unity. The sensuous charismatic person is filled with the gifts of the gods, but may also be the recipient of the projections of the many who do not honour their own.

The charismatic person is usually seductive without necessarily intending to be. He or she lives by his or her own values, rules and codes, not those of others, nor those necessarily created by society. The charismatic person is often a conventional non-conformist, defiantly primeval yet cultured and wise. There is something disturbingly sexually ambiguous and outrageously sensual. The world soul-sense, the *anima mundi*, finds expression through this personality; there is a sense of timelessness and diversity. Psychology is not enough, for here is mystery at work.

Every human being has grace given by the gods, but it is dished out both randomly, and unequally. There is no judgement in nature, nor in the world of the gods. Some of us say 'It's unnatural to be so charismatic! How can she/he be so radiant, so erotically glamorous, so mystifyingly sensual?' But the unnatural *is* nature. Nature is unnatural. It's also more likely that those utterances are concerned with 'how dare she/he be so glamorous', and 'why not me?'. Glamour and charisma are evocative of the mystery of the senses, but how conscious or aware we are of this secret force within ourselves depends very much on our personal world view. Then, of course, there are those who manipulate and abuse charisma for the purposes of power.

Amorality

Glamour is the mysterious gift which is totally self-governing. It looks out from the glamorous person and enchants others to itself. This quality bewitches and deludes, it is the unity of Aphrodite's erotic aura and Hermes' protean lawlessness. The glamorous person carries an enticement. Fire and water are invoked, and an alchemical burning of two people in passionate self-love stares out from one face. This gift is spell-binding to others. In itself this quality is neither containable or controllable, spirited and numinous glamour pervades where it choses. It incites and burns outwardly, and sexually in flux it draws chaos inwards and creates an illusion of self-will and self-confidence. Glamour, like love, enters into the human organism, a-moral. We in turn feed off both love and glamour and make them moral or immoral as our other senses awaken to the mystery of both. We may suffer for loving, we may cherish a moment of sensual pleasure for loving, we may live in hope for glamour, and we seize hold of it when it comes to us through another human being. But glamour, like love is that which cannot be contained, and it sometimes forces us into becoming either Narcissus or Echo.

Echo or Narcissus?

Echo was cursed by Hera to only ever be able to repeat the last words she heard others say. Poor Echo fell in love with the self-centred, effeminate and auto-erotic Narcissus. Every word Narcissus spoke, Echo imitated until he rejected her. With no voice of her own, Echo could only reflect the shallowness of Narcissus's egotism. His self-love eventually led to his own fate when Nemesis cursed him to fall in love with his reflected image in a pool which he could never leave. Echo eventually 'echoed' away, her imitative voice heard only in nature.

Glamour can enchant us into being Narcissistic if we are not careful, and it can also make us Echo-like. Like Echo, we may imitate the best, we may fashion ourselves after the famed and the fortuned, or dress ourselves with the imagery of another's charisma. We may try to live in or through the image of a partner, lover, idol or parent, denying our own inner qualities and needs. Like Echo, we may imitate the person who only values his or her self as worthy of love. Echo was once the consort of Pan, yet her curse meant she was disconnected from the wildness, the beauty and the inexpressible body of nature. Her only way to rejoin the pantheistic world was to be reabsorbed into it through her rejection. Likewise, from our individual viewpoint, we often see glamour and mystery solely as the business which promotes it as a commodity. We may get drawn into glamour to be an echo for someone else. Alternatively, we may only ever be a poor imitation, not knowing how to take responsibility for who we are, not realizing we have our own personal 'echo-less' nature. There will always be as many Echos as there are Narcissi.

Those who are like Narcissus will promote their own self-love, they may use glamour to manipulate, they may fixate on the mirror, denying others the chance for reflection. The glamorous Narcissus loves only him or herself, she or he must shine better, must polish his own looking glass to perfection. To love yourself more than anyone else could ever love you, means you are safe from the very rejection which as Narcissus you may bestow on others.

Oscar Wilde wrote, 'Wickedness is a myth invented by good people to account for the curious attractiveness of others'.[2] Glamorous, attractive people usually draw both men and women to them. This is why it is easy to become a living icon, a fiercely ruthless politician, a benevolent leader of civilization, an enchanting, seductive film star, an opera singer or a guru. Every 'wickedness' one denies in oneself becomes embodied in the glamorous attractive person. Every unconscious vice and virtue is polarized. We stick 'post-its' and labels on the charismatic person as a repository for our own shadowy nature, unless we at least try to understand what grace and mystery are all about. Glamour is about the unusual, the different and the strange, because the quality is itself primeval, not ordered and discriminate. Nature's life-force scintillates and seduces through the process of human relationships and the magic of charisma. But if the mysterious becomes locked in a sense of physical intensity or the magic becomes imprisoned in humanity's will to power, it seeks escape. Excessive glamour is self-destructive, and deadly to others. Power becomes glamour's only 'Achilles heel'. Like gutted fish, the once-glamorous who have fallen from grace or power have sullied scales; no longer do they have shimmering eyes.

Erotic Glamour

Erotic and sensual charisma is Aphrodite's gift. Of itself it is genderless. Why is it then that our collective imagination has largely attributed this quality to women? Women are the most basic symbols of the feminine or yin-sense. Historically, we have ingrained images of the feminine, of body, of beauty, of moon, of the sea, of night, all these symbols have left us bereft of the fact that men and women are just creatures. We are created, and our one huge difference is a biological one, but it is a difference which has become amplified as we charm and work the spell of nature. Once we began to see differences, once we see woman as egg and man as sperm, woman as vulva and man as phallus, we have immediately split ourselves. We know man is built with a different muscular system, and is physically stronger. Then the track widens, the tracker finds new scents: woman lactates and menstruates, man merely ejaculates. Men hunt out the eggs and the caves which hold the eggs, woman is the cave, surrounded by water. She is the guardian of the underwater cavern and she must entice man into this cavern. Charisma seems to work through those who honour both the masculine and feminine within, or rather those who acknowledge the physical and biological differences of man and woman but don't erect a fence between the qualities of yin and yang. But could man and woman have ever been one erotic glamorous fusion themselves?

The Greek playwright Aristophanes told a story of how the first beings were androgynous. These hermaphroditic balls had four arms, four legs and two heads. They existed in a state of complete sexual bliss, harmony and autonomy. The gods, however, were convinced they shouldn't be having so much fun, so they cut them in half. One half was doomed to spend the rest of its existence searching for the other half to make itself feel complete. The charismatic person usually feels already complete, like this androgynous being, or the Djanggawul or World Egg. The person who is innately charismatic has the power of eroticizing love. The rest of mankind is still running around looking for the lost half in the opposite sex, and sometimes in the same sex, but the truly charismatic person knows it to be him or herself.

The sensually glamorous woman has gathered a collective bale of criticism, envy, hatred and fear in her haystack. Passion and mystery mobilize more vividly in woman simply because she is a cyclic creature, a personification of all that we perceive as feminine. Camille Paglia suggests that 'glamour is a gift under no one's control. In women it may peak and ebb with the menstrual cycle.'[3] Cleopatra, as already woven into our story, was perceived as exotic, erotic and beguiling. She retained an eternal image of the femme fatale, seductress, an Eastern Yin-cess to Western princes, an archetypal image of the glamorous, powerful woman. Nature-culture in Everywoman shape-shifts with her changing rhythms. Her outer expression of erotic energy and glamorous aura pervades as fascination, electricity and seduction. Her rhythms are attuned and vibrate to the earth, to the cycles of nature and the dialogue of the macrocosmic nature and microcosmic culture. For the Victorian poet Arthur Symons, Cleopatra 'could draw the stars out of the sky with love'. The enigmatic woman can also draw the waters of the ocean around her like a diaphanous cloak.

Protean Nature

Proteus was the god of the sea, eternally absorbed and transformed through the waters of the ocean. Sensual glamour is fluid, transforming and liable to rise and fall with the tides and the moon.

Woman, her body closely aligned to these natural rhythms of time, is more adept and accomplished in the world of mystery. She, like Proteus, gender aside, flips her coin and changes with the tide. Her glamour surfs every wave, then dives deeply to the pearls of the ocean floor. She is less troubled by glamour's secret for her body is designed to invite by mystery. The biological purpose of man is to be powerful, hunt and generate. Filled with 'desire', he may well resort to attempting to manipulate this self-regulating energy simply because it seems so alien to his own biological persona. Fluidity, magic and desire have become ingrained in his mind as feminine qualities, which he then sees only reflected in woman. The purpose of his body-sense is not protean, it is projection.

Charles Manson has a powerful magnetism, alluring, hypnotic, fuelled by family curses, social allegories and an inferiority complex, the outer world of society's glamour beckoned, and his inner world of glamour listened to the call. Charismatic gifts fuelled and intensified Manson's darkness, for the daemons of primeval power feed furiously on those who do not honour them. The destruction of others to feel self-love is a rare manifestation of glamour's power, but when something so mysterious is sustained by the will to omnipotence then personality is irrelevant, and the forces themselves invade the ego and take over.

Glamour and Power

The instinctive charismatic person has access to a magnificent power, simply because she or he can seduce some or all of the people, like the Hitlers and Ghengis Khans of this world; or delight and fascinate, scintillate and enchant like Maria Callas, Oscar Wilde and Tina Turner. The intensity of glamour's force can bring you a Mephistophelean nightmare or a *Midsummer's Night Dream*.

There are many women who are born into a position of power, like Elizabeth I, Mary Queen of Scots and Queen Victoria. Certainly these women exuded a presence because their power was absolute.

They had no choice but to express a charismatic quality, to entertain the gifts of the gods and live out their destiny. To be born into glamour, or glamorizing people with the intention of exploiting that glamour, is not the same as those who carry destiny's grace and are obliged to channel the expression of it. This is where the world of art, music, literature, politics and passion are so valid. Today, the world of 'celebrities' welcomes eccentric presence and extreme or intense characters into its lair, but the magical aura around you may be frozen in media ice if you are not prepared to pay a high price.

Madame de Pompadour

Madame de Pompadour was the alluring mistress of Louis XV. Not born into nobility, she had considerable influence over the King, particularly in his passion for art and architecture, and it was said that she was discreetly significant behind the scenes in French politics. Her charisma was not an aura of dramatic ostentation, or sexual ambivalence, but apparently of an elegant fragility. A silver spangled silkiness rather than an underdressed sex kitten of the tabloid kind. By the age of 20, Jeanne was the centre of a circle of artists and intellectuals including Voltaire, her beauty renowned, and her presence stupefying. Louis brought her to court as his mistress, and her brilliant mind, her grace and her gentle glamour were enough to keep her permanently as his courtesan. Later however, he refused to indulge in her intellectual circle known as 'Les Philosophes', which was to become one of the contributing critics of the monarchy and its subsequent downfall.

Isadora and Garbo

Isadora Duncan was noted for her unconventional dancing and her unorthodox lifestyle, which were both challenging and provocative to late Victorian values and morality. She had stormy love affairs with stage designer Gordon Craig and millionaire Paris Singer, and bore two children, but carried on her free-spirited style which, pagan in its content, was more expressive of human nature and the senses of the body than the contemporary formalization of ballet. Isadora danced barefoot, wearing only a Greek style chiton, a loose voluminous tunic which was revealed most of her body. During the 1900s she danced her enigmatic and sensual whirl through Paris, London and New York. Her life was filled with tragedy – glamorous women are also often fatal *femmes*. After the death of her children in a drowning accident, she married the poet Sergei Esenin, but they soon separated and Esenin committed suicide in 1925. She returned to France and was strangled in Nice when the scarf she was wearing became tangled in the wheels of her open-top sports car.

Garbo was a different category of charismatic woman, a talented aloof actress. This is a profession where charisma steams into gutsy fantasy, if the imagination is boundless. Glamour lures many to the stage and many of those who dare are those who honour their intensely passionate or powerful personalities. Greta Garbo became one of the legends of the cinema. Her magnetic aura and icy distance gave her a reputation as the Swedish Sphinx. Her distinctive, deeply resonant voice was considered seductive and erotic. As Anna Karenina, Queen Christina and Camille she acquired worldwide success, but after unfavourable reviews for *Two-Faced Woman* she spent the rest of her life in seclusion. 'I want to be alone', her famous line from the film *Grand Hotel*, was her most profound statement about herself. Lonely? Loneliness can be powerful, it can be a positive quality for the auto-erotic and intensely self-generating person. Glamour, like love, may prefer the lonely to the distractions of those who do not understand its mystery.

Flowing Hair and Glamour

One symbolic manifestation of glamour is hair. Hair is floating, combed, wigged, braided, plaited, hennaed and fashioned according to the cultural expression of the time. Yet it is an ancient symbol of the mysterious glamour of woman and also of the mystical world itself. By the end of the nineteenth century long golden hair, once fashionably coiffeured and styled lavishly into ringlets, became a fearsome temptation to men. Hair was viewed as a sign of woman's intellectual weakness, although it may have been a backlash against the collective confusion around sensual pleasure. Sexual confusion and contradiction was a deep thorn in the traditional armoury of the era. There was a collective anxiety about hair as the Gorgonian reprisal sent to trap men from their intellectual wisdom and dominance. Hair, particularly in art and literature, still represented the serpent. Medusa lived in every woman, glamorous, sensual and powerful. The longer or the thicker your hair the more you were likely to be dangerous to a man. Even today, whatever the fashion, long tresses or braided hair denote something archaically enticing and alluring. Long hair swirls around a body, and swirls around both bodies when joined in sensual pleasure. The sexual connotations were problematic for the *fin de siècle* artists and poets, but it was the fear of their own masculine potency that forced them to denigrate the hair of woman rather than to enjoy it. Swinburne's *Atalanta in Calydon* reminds us:

'Not fire nor iron and the wide-mouthed wars
Are deadlier than her lips or braided hair.
For of the one comes poison, and a curse
Falls from the other and burns the lives of men.'[4]

Arthur Symons too comments on the fearsome demonic glamour of a woman's tresses: 'She has bound my neck within the noose of her long hairs, and bound my soul within the halter of my dreams'.[5]

Earlier, Heathcliffe and Cathy in Emily Bronte's *Wuthering Heights* were two brandishing torches of hair, lips and violent desire caught and held in a whirl-flame of their own making. Their glamour was spellbinding because it was the androgynous glamour of one seen as two. Anger, tempestuous desire and passion, thirst for sensual power and sexual gratification that became remembered as an allegory of Victorian glamour.

Hair has always been associated with paganism. St Paul feared the spirits which women unleashed when they let down their hair, and ordered women to cover their heads 'because of the angels'.[6] Even today, women are required to cover their heads in many religions lest they invite evil spirits or 'let loose' mystery and pagan powers into churches or places of worship. There is even an old belief that bats get entangled in women's hair because its length and thickness has the power of spellbinding anything to it.

In the East, Tantric mystics believed that the cosmic energies of creation and destruction, of birth, death and regeneration were unleashed with the unbinding or binding of a woman's braids. Goddesses and women alike were responsible for the weather and the spirit world, and all three forces of nature seemed tied up in a woman's hair. Isis, Kali and Athene were supposedly able to control the fate of man depending on whether they plaited or unbound their hair. The longer the hair the more glamorous the image. One of the earliest sexual allusions to hair was when the Egyptian goddess Isis restored the life spirit to Osiris so that he could be reincarnated as Horus. Many myths stated that Isis coupled with Osiris while he was dead. By stirring his phallus to life, she covered him with her hair and warmed him temporarily into being. She 'produced warmth from her hair, she caused air to come . . . She caused movement to take place in what was inert in the Still Heart, she drew essence from him', i.e., his semen.

Apart from being associated with witches, gypsies and Eastern exotics, hair is one of the most glamorous attributes of a woman. Glamorous, in other words mysterious, it became feared and condemned by traditional societies as imparting pagan and mystical power.

Glamour seduces those who need enchanting. Heavy fines or penalties are the result of believing one is above enchantment. For the needy never know they are needy until need becomes hunger, and hunger becomes addictive obsession. Thirsting for enchantment, many men and women are drawn to glamorous icons or seductive mirages. But enchantment is nature's gambit. It forces us to confront relationships as the source of disenchantment, so that we may be enchanted again, to work out and weave through our fate. Glamorous enchantment is earth-bound. It forces you to look down to the pebbles beneath your feet which will one day be sand, and fills you with the strangers of yourself which you have yet to meet.

Smelling Violets in the Grass

The sensual challenges the mind because it implies all that is not of the mind, the sights, sounds and fragrances of life which are the catalysts for the mind. Do we imagine the senses into being, or does imagination discover us through the senses?

Eastern philosophy is both imagination and the sensual, as yin and yang, as Tao, as ch'i. Even the Eastern philosophies and beliefs are perceived as 'feminine'. The esoteric Eastern sensual delights are an alluring mystery to many who have found the traditional philosophies and religions of Western civilization rigid, structured and bound mostly by evidence, fact and measurement. The mundane earthiness of life is replaced by the erotic earthiness of life. Glamour seems almost to be alive, awake, tangible and solid. It has purpose, and the more exotic and foreign the concept or image, the more mysterious it becomes. Sensual glamour however seems to bridge both East and West. Women like Jemima Khan and Queen Noor of Jordan are Western women cloaked in Eastern glamour. Noor, a powerful humanitarian figure among the Jordanian people, is also a woman of grace and sensual presence. Poets, troubadours, astrologers, philosophers and crusaders journeyed East to discover the glamour of the exotic, and it is through the embrace of Eastern soul-sense and Western mind-sense that something inexpressible is constellated.

The poet Keats once commented on an Anglo-Indian girl at a tea party in 1819: 'when she comes into the room she makes an impression the same as the Beauty of a Leopardess. . . I should like her to ruin me'.[7] A modern analogy would be similar, but 'he' would probably not think the last thought of Keats, not consciously anyway.

She may be flattered by his thoughts if she could hear them, equally she may be insulted or reject his self-abnegation, or her glamorous image may be used as a defensive attack. But this girl would also bridge East and West, the hauntings of all her inheritance, her primeval ghosts would filter through the sieve like an organic Darjeeling tea laced with Cointreau.

He asks, 'Did you smell the violets in the grass? Did you put your nose to the ground and inhale the sweetness before you came inside?'

She smiles back. 'Yes, You must be a poet.'

'I try, and you?'

'I do not try, but I am a poet'.

A modern woman may be more aware of who she is than the Leopardess girl of Keats's muse, but her sense of him will be the same. Like Lady Caroline Lamb's tempestuous affair with Lord Byron, her dark desire rejected, her monstrous passion directed ruthlessly to destroy herself and others, she mercilessly self-sabotaged her own glamour and charisma. She too knew what poetry is.

Poetry is Eastern luminosity splattered with the sensual decadence of the West. Poets smell the violets in the grass; they sense the fragrances of destructive passion, of love, of fear; they sniff out the qualities of the inexpressible and try to express them. They gather images, ideas and senses, and yet there is a sense that poets are filled with the ready-gathered. Glamour, poetry and charisma are 'incognita's' threshold, this is the liminal point, the cusp between the known and the unknown, and this is where the sensual truly lives.

PLEASURE HUNT

NOTES

1. Shelley, v. xii, 137-9
2. ed. Edwards, p.223
3. Paglia, p.376
4. Swinburne, p.265
5. Symons, p.58
6. 1 Corinthians 11:10
7. Quoted in Praz, p.285

Chapter Seventeen

Sensually Yours

'You could not discover the limits of the soul, even if you travelled every road to do so; such is the depth of its meaning.' Heraclitus.

We have a conditioned reflex to the word 'sensuality'. Even though you may have read all the words in this book, the cultural matrix suckles us with illusions of what sensuality is and isn't. Escape from the chains of our locked-in perceptions is not an easy liberation. The word 'sensuality' suggests decadence, sexuality, *femmes fatales* and glamour most feminine. The sensual certainly encompasses all these things, but it has also a more profound 'sense value', and if we haven't got the message yet or refuse to uncloud our minds, then we never will. If we are honest, our rigid thinking and collective assumptions are manifestations of the dilemma of ego-centred living and the basic fact that reality is subjective. We can only express and perceive radical ideas or new dreams if the radical dreams and fresh thoughts chose to come into us.

But this says nothing about how we feel about our personal senses. Often biased towards the common view, we don't value or understand our personal inner needs. We may become entranced by family myth because it seems to reflect our own deepest intent, the romancing of our senses by traditional values may simply be due to an inability to clarify our own, or we are fuelled by outer motivations rather than inner integrity. Perceiving sensuality as wicked or seductive is fair as long as we know why we are perceiving it this way.

We can only ever be as physically sensual as our bodies can be. We are bound by the limits of our flesh and bones. But the sense of mystery, of intuition, of sensing one's own being is boundless. The imagination is our most fundamental connection to both the sacred and the profane. It is our imagined world which can lead us into sensual awareness we have never experienced before. By honouring our bodies and taking pleasure in our senses we are invoking all the gods and nature's necessity. Fate becomes our free will, and free will becomes our fate. We are ironed on a board that requires we take responsibility for our lives, but if we don't spray the rich cloth of our being with water's mist, we become wrinkled, creased, crisp and rigid in our thinking and in our feeling.

Those who attend to and take pleasure in their bodies, or at least respect the body's ambiguities, are also eroticizing nature. The sensual person keeps nature's fingers playing the strings of our eternal violin. Without passion, sensual love and erotic desire we become hardened, a solidified place where soul is unwelcome. Without disharmony there would be no harmony, order relies on chaos to keep the images changing, to keep moving with imagination, which is why there are probably as many sensual types as there are people.

Sensual 'Types'

To be aware of the sensual nature in us does not mean we have to be hedonists. Hedonism is derived from a Greek word meaning pleasure and sweet. The doctrine of hedonism, in philosophical ethics, was the belief that pleasure was for the highest good. Hedonism therefore can involve anything that gives pleasure, but unlike sensuality, it evokes a quest to acquire the highest good through pleasure. The sensual does not rely on a hierarchical need, it is innately given and pleasure and pain are both sensual experiences. In each of us is the ability to be aware of one's sensual nature, to differentiate between what feels good, and what feels not so good , but also to have the sense to 'treat those two imposters just the same'. There is no seeking of the 'highest' good, it is acceptance of the baseness of being animal, and of nature's ambiguous unpredictability. Sexual sensuality is a natural and entirely subjective response and instinct, whereas hedonism implies taking pleasure for its own sake. Instinctual love is the vagabond of the body-soul who comes to us begging.

For most of us, these instinctual pleasures of the body are mostly found through our sexuality. Desire burns us, fire lights us and we rage and burn until we fall into the bonfire of mutual passion. Then comes the waters of ecstasy and unworldliness, the profound and dark erotic earthliness of emotion and feeling, the mind stretched beyond all possible thought and into silence, or disbelief, or desire again. We are slaves to our bodies and our desires, and we are slaves to the gods and each other.

So are there different types of sensual people? Four basic elemental forces seem to align with the archetypal energies which form our expressions of sensuality. The Renaissance philosophers defined these basic temperaments as choleric (Fiery), melancholic (Earthy), sanguine (Airy) and phlegmatic (Watery). Earlier cultures used similar symbols, and the four elements were fundamental principles or energies which the ancient Greeks believed made up the universe. Desire forces us into finding ways of channelling its expression, usually through the physical senses. We touch or tease, gaze or freeze. Each of us has our own pattern, styles of sensual expression, especially in the arms of another. Here are some.

As sensuality has been allocated a 'feminine' quality, the Greek goddesses personify the basic personalities and expressions of the four elements. These modes and transference of expression are applicable to both men and women, but will diversify depending on individual psychology. Incorporating different aspects of each of the elemental types means our personal sensual expression may change and shift according to whom we are in relationship with and what we are projecting or disowning in ourselves. Reading these descriptions may invoke a response or a reaction in us, if something 'pushes our buttons' then it is certain to have some meaning or underlying significance to us. We may need to look to where the button gets stuck.

Fire-Sense

Circe, witch and temptress, was the daughter of the cyclic sun-god Helios. She lived alone on the island of Aeaea, bewitching those who landed there and turning them to swine. She was reformed by Odysseus with whom she fell in love. Her earlier history shows her implacable nature. Jealous and volatile, she first fell in love with Glaucus who was already entranced by the nymph Scylla. In a fit of jealousy, Circe poisoned Syclla's pool and the nymph was turned into a hideous monster, doomed to lurk for ever along the cliffs above the straits of Sicily.

Fire-sense people have dramatic senses like Circe, and learn to lure and seduce through their intensely sexual aura. These people are the wild-cats of the night, more likely to be attracted by the power of sight, to search among the many, rather than test out the known few. Fire-sense tracks and smells out its prey, rather than being hunted by others. Stalkers supreme, they know intuitively, another fire-sense, the wisdom of relationships and of the Delphic oracle. Fire sensualists take pleasure in the chase, the first touch, the responses of suggestion and of the imagination. They smell keenly the wind of change, any new essence must be assimilated voraciously. Ideas and images flow freely in every embrace. Fantasy and pleasure must be indulged as adventure and a quest. The reality of dirty laundry and the earthier sensuality of food, furs and lace does not particularly motivate them. She or he prefers the tactile sense to be that which touches possibility and probability, not here and now, not mud-wrestling or a bed of thorns. These fiery sensualists are passionate and proud, and cultivate their beauty. Extravagance and outrageous sensual indulgence, however, may be a pretence. Denial of their own neediness means they often avoid deep emotion or the feeling senses being evoked in their relationships. They are sexually lonely, but can turn their lovers into slaves, like Circe's swine, simply for self-gratification, rather than mutual pleasure.

Atalanta personifies another fire-sensualist. These are people who want impulsive touch; they are sparked, fast, flamed into passion As quick as Atalanta they are spontaneous and burn quickly. Slow languid pleasure may bore them. Tomorrow's pleasure is always more intoxicating and sensually stirring than today's. The senses are more tuned to winning and hunting. They are hard to satisfy and rarely easy to predict. Artists and poets, writers and musicians, the Gustave Moreaus, Isadora Duncans and real-life Scarlett O'Haras are all fire-burners. They are people who are lovers of passion, who respond to the imagined, to a sense which provokes notions, or an image to which they can relate. Words, pictures, sounds, smells all are necessary to energize imagination, but these senses must be transformed into inner fantasy, the dreams and visions of fiery sensualism are then thrown back into the cauldron of passion for re-burn. Phoenix aflame in a cardboard box.

Earth-Sense

Aphrodite's most exotic seduction fuels and fallows the earth-sense personality. Fecund and physically strong, inscrutable and richly powerful, these are people who are intensely earthly, yet often fearful of their own sensual nature. Earth-sense requires the tactile senses to be evoked for maximum pleasure. Carnal contortions are rich qualities, desired and given. She or he may identify with Aphrodite's most ambiguous beauty, the ominous resplendent princess of love, and the dark power of her enticement. Quietly, this is a sensualist who looks in mirrors of polished steel and has flashing eyes that watch and wait. The violent, chthonic world of Lucretia Borgia or the earthly erotic writing of Anaïs Nin may be expressions of earth-sense personified.

Aphrodite beckons and those to whom she smiles must follow. Earth-sense thrives in the moment of here and now, of being in the place and experiencing their physical senses more than any other 'type'. This innate awareness of the present means they can forget the past and rarely think to the future. The sense of touch, of seeing, or hearing and thinking can be observed, watched. Desire and its envoys may find free expression through the sensual messages these people can silently seem to convey to others. This is the person for whom *epithemia* is not to be feared. Remember, *epithemia* is the basic instinctual need for human touch and contact. Tangible love, physical action, the fragrances of sex, the diversity of nature, all must be earthed and grounded in the body.

Aphrodite's possessive and vengeful nature crystallizes in the earth-sense person when passion turns from physical sensuality to emotional sensationalism. Earth-sense envies others their imagined world of the senses because earth is irrevocably attached to living in the world which seems unimagined. Bacchus and Pan, hedonistic, carnal and sexually active, are both interlopers in their interworld. Earth-sense does not require erotic and profound intense suffering, but it does require complete fusion of bodies. These people do not feed on the emotion and feeling senses, but yearn for the smells, the taste and sight of sexual enjoyment. The body is the shrine, an origamic sensual unfolding, rhythmical, cyclical and physically committed.

Pasiphae is another darker personification of earth-sense. Cursed by Poseidon, the Sea God, with a desire to be ravished by the white bull her husband King Minos refused to release, her reputation for sexual bestiality was unrivalled. She gave birth to the famous Minotaur, and later ensured her own revenge on Minos, so that whenever he copulated he ejaculated poisonous vipers, insects and scorpions to destroy his lovers. Earth sensuality can be possessive and resort to sexual control. This person may unconsciously choose lovers or partners who retaliate with poisonous words. Jealousy is common when others play into the field, because physical love and the sexual senses are the most fundamental validation of self-worth. Earth-sense can wound before it is hurt, it has a gleaming knife of silver beneath a placid velvet cloak.

Air-Sense

The power of the voice, words, language, the beauty of communication between two people are air-sense rapture. Tongues speak and having spoken move on to a fresh landscape of words. Poetry is the body, and the body is poetry. Air-sense gathers thoughts, dreams thoughts, aspires to dream rather than to indulge in physical sensuality. Air-sense is ambiguous, often imagines bisexuality and woos diversity. The interplay of language, sexuality and idealism becomes invaluable. Hermaphroditus, the offspring of Hermes and Aphrodite, only truly became dual-sexed when he bathed in the river and the water-nymph, Salmacis desired him, merging with him sexually forever. The sense of androgyny, of equal balance and relationship, lead Air towards many sense encounters. Air is the most likely sensual type to enjoy both man and woman as sexual images or fantasies. Air-sense is more aware of the fusion of minds, of souls and of bodies, they rework, revision and play with the exchange of the senses. It is the quality of connections, of senses merging with mind, with mind merging with body, that gives air clarity of vision.

Elizabeth I and Joan of Arc breathe in our stale air and breathe out sensual air. The former a sensualist of the mind, of intellect and passion for living, the latter like a mystical embrace with a loaded air-gun when burning on her stake, air shoots flames into her melting arms. Both women immersed in connection-awareness, but of a different dialectic.

Often air-sense finds it difficult to identify with 'feminine' images, yet the primitive quality of the sense of the inexpressible, of words as gods, is unique to air. In fact, Air can sense the sameness of masculine and feminine so inextricably that these people do not perceive the world in opposites, only in wholeness.

Nyx was the personification of night. She was both winged, awesome and subduing, and carried her cloak of stars, black veils covered in twinkling diamonds, eternally unseamed. When the air-sense aligns to the ancient power of Nyx, he or she can embrace the intuitive threads of ancient wisdom, the sixth sense of uncanny knowing. Because air people may fear the undercurrents of the feeling senses, they automatically close down their emotional natures in intimate relationships. They may cloak themselves in a false image to protect their low self-esteem, and indulge in manipulative power games behind the false persona of compliance and calm common sense. If air-sense does not fly into the realms of both the rational and the imagination, if it does not honour and identify its desire for passionate communication, or intimate exchange of sense messages, then it may resort to criticism and impossible expectations. The idealistic notions of air are as diverse as their sense of dispersion; ether and breath, the winds of change and the rush from the satyr's hoofs. Air swoons in senses of opaque clarity.

Water-Sense

Lilith is one of the the personifications of water-sensuality. As handmaiden to the goddess Inanna she sets out to enchant men into initiation in the temple. But she only took those who were to be healed or transformed by sexual arts and ritual, awakened into the sacred through the profane. Water-sense incarnates as enchanters, beguilers. The wild fluidity of the sirens, the haunting songs of the mermaids, are heard by water-sense people who have an uncanny sense of deep awareness. They feel their feelings, they feel their senses, they know their inner beauty. Water-sensualists may break taboos, plunge deeper into emotional and erotic love, they are drawn into intimacy and out of it over and over again. The body is made up mainly of water, and the rhythms of woman's body are like ocean currents. Water-sense feels circled by the waters of life, whether male or female. The darker realms of the feelings are either intensely sensed or disowned. The senses of Water are sea-salted, not supersensate as dolphins or whales, but coral-formed, ancient cavernous depths of knowing. The primitive life-force is easily manifest in these people. They are sensitive to others moods, needs and feelings, which they can believe are their own. Louise Colet, Flaubert's beautiful, wild, passionate mistress, may have been his muse or inspiration for his novel Madame Bovary. The scandalous account of Louise Colet's liaisons, written by herself, reflects her own intense, passionate, sense of love and life.

Sexually water-sense merges with others to embrace the divine in everything. But water-sense may flow like a mountain stream when there is no rain, identifying with Eos, the goddess of the dawn. There is intuitive, mystical, iridescent pleasure on a spider's dewy web at dawn.

Then there are those who sacrifice their own sensual need for the other. Calypso fell in love with Odysseus. She entranced and captivated him under a magic spell, refusing to believe he was already tied to another woman across the seas in Ithica. Hermes was sent by the gods to tell Calypso to relinquish her spell on Odysseus. Reluctantly, he abandoned her and she was left alone. Water-sense seeks to avoid the truth, like Calypso. The harsh facts of physical violence, ugliness, the horrors of mankind and individual greed, can induce a yearning for a lonely island, only to take pleasure with one who may stay awhile, always knowing they may leave. But if water-sense feels feeling, instead of only aloneness, then erotic relationship becomes a powerful channel for the deeply exotic richness of their water warmth. Water-sense is elusive, gentle, surging, passive, but equally Shiva-like, as intense and compelling as Kali, Medusa and Lilith. This is the dark and dangerous sensuality of smothered passion. The pillow of golden rain pressed on a face of etched glass.

What Are Our Fundamental Needs?

The body is insecure. It is not to be trusted, for we know it will die one day. The notions and images we perceive from our ego world which believes itself to be locked away in the most fragile of armoury, the body, are the most fundamental energies which enable us to open up to the life-force within. Fate, destiny, death and mystery both inspire us and shock us. The sensual is our instinctive base nature. Our outer values and needs are different from person to person, but they are based on the rawness of mortality and the fundamental need to be loved and to love. Trusting in our senses we may discover a changing awareness not only of ourselves but the beginning of understanding, a glimmer of compassion towards others too. If our minds are unclouded by body anxiety we can honour imagination and soul.

Our own sensual world is a very private place. We may take pleasure in the darkness, the instinctual, the body tattoos of life. We may live it as an illusion, cloaked in adornment, beautified according to fashion. We may abstract our senses, blame others for feeling, emotion and sensations that are in truth our own. We can disclaim all our senses because they are the root of our flaws, virtues and shadows. By looking at those qualities we admire or envy in others, by noticing who and what 'gets' to us, or pushes our buttons, to have the sense of this awareness, means we can begin to acknowledge our own personal sense values. For example, if compliance is in our nature, then being aware of a sense of compliance rather than resenting those to whom we always make compromises means we may begin to acknowledge our own self-worth rather than doubting it.

What Are We Hunting For?

We are the dark riders of the soul. We are Pan's venerators, and Dionysus's drinkers. We are born predisposed to sense out love and hate, to feel and to be filled with desire, to hunger and to yearn. The qualities are the same for us all, but it is our unique perception, formed and crystallized out of star-dust, that means we may think rigidly where others think malleably, we may love endlessly, while others love only self, we may bear pain, while others bear only resentment. We are all hunting for something through our senses. But what is it exactly that we are hunting for? Biologically, we are hunting to continue the species, socially, we are hunting for the good of the species, but personally we are hunting for soul. We sniff the air for good news, smell a rat, or get a whiff of hatred, envy or blame. We sight another body that may provide us with love and physical pleasure, we are filled with the sense of desire so that we may express our intentions, but we are always looking for soul.

There are always those who are drawn waywardly, outlandishly, to passion, pain, to intense erotic relationships sensually enticing, always inexpressible except through sexual love. Desire is sacred, it moves in with us when least expected. It is a phial, a love potion sewn into Zeus's thigh with Dionysus.

Then there are those who are always drawn to the numinous, or idealistic relationships where the separation from a sense of meaning seems as valid as the need to merge with another to find it. We hunt for pleasure in different ways. We acquire the sense of pleasing ourselves through different means. Some of us have simple needs, others more complex, but whatever our personal 'taste' , whatever it is that sends us, or arouses us or touches us, we are searching for a meaning to our beingness, the sensual that is nature's indelible soul.

Sensual Power

Power can be the most destructive and the most creative force in human relationships. Whatever the underlying personal motivation for those who must have power at any cost, when we rely on our sensual nature to acquire this sense of omnipotence, we are exploiting instinct through charisma.

We cannot have thought or feeling without the senses. Before we can have an image of a rose in our mind we must see it, touch it, feel it, smell it. Before we can have a notion or fantasy of sensual pleasure we must taste, see, touch and smell to know what is pleasurable to us and what is not. We cannot have a concept unless we have a sense of what a concept is. We cannot have anger unless our own perception of anger has been a personal reaction, or a sensing out of what brings feelings arising within us. Once we abuse that sensual world, we begin to feed off the sense of power we have over others.

A sense of power commonly manifests itself in sexual wars, love triangles, mistress versus wife, or wife versus another wife. Betrayal, vengeance, jealousy and suffering are our personal *grimoires*, our magic-spell books, and each one is a clue to making conscious the shadowlands of our mysterious being. We search desperately for the perfect relationship, or remain in static, sticks and stones double acts, inconspicuously creating resentment, while fighting for power. Books tell you how to keep your relationship alive, how to save your marriage, how to understand your 'man' or 'woman'. These are valuable, but only if we realize that our instincts often over-ride reason.

The tradition of marriage becomes an illusion only because we have forgotten that marriage is an enterprise, a structure, a binding and a contract. And mostly it is not necessarily inclusive of passion, love, sexual romance, and rarely spiritual grace. Love is a mystery to us all, and works through us differently. It moves in if it so desires and out again at will. Sensual love is not fixed by legality or binding promises, it is the burning, the blackening, the reddening of the alchemist's fire. Sexual relationships don't have to include marriage, and they don't have to include love. Sex is not love, but the sensual incarnates most freely in those who can learn to honour both unconditional love, and Eros, the binding love of all the senses, as one and the same.

Last Words

'But I who am the eternal woman, the archetypal feminine, I do not speak to the surface of consciousness, the sophisticated mind that the novelties catch, but to the archaic and primordial that is the soul of every man, and I will pit my charm against that of any fashionable woman. They may have lovers, but I have been loved.'[1]

Dangerous or Endangered

The sensual and woman are not dissimilar, for the sensual is perceived as being a feminine image, and the feminine itself is symbolic of that which is not masculine. They are only exclusive to one another because we have chosen them to be so. If the sensual displays itself through women more readily it is only because the femaleness of women amplifies the feminine principle and thus casts woman in a doubly exaggerated embodiment of all those qualities. The female body enhances the sensual, it adds the mystery of nature's secret place, it lustres and glazes our perception of what is feminine, it provokes us to carry the idea that woman *is* sensuality.

All the archetypal images we have of yin and yang result from our seeing woman and man as prime examples of the polarity of the tension of opposites. Yet the sensual itself is genderless. This quality manifests as mystery, glamour and charisma, it is intimate partners with Eros, Aphrodite and Dionysus, with earth, nature and the pagan world.

Throughout this book we have seen how sensuality has become distorted into being perceived as merely wantonness, lasciviousness or self-indulgent hedonism, and how we have assigned subjective connotations and qualities to the sensual. We have also seen how the sensual merged with our sense of sexuality, becoming incarnate as only one sense, that of the sense of physical pleasure. Sensual sexual pleasure is to enjoy self and the other, it is to find what delights and excites you, it is to have a sense of being rather than a sense of apathy. This is the greatest sense of being we can have and is an interface for the sacred and the profane. In itself, sensual sexual pleasure is not dangerous, it is only those who exploit its power and corrupt its qualities that debase and demean the sexual senses.

But the sensual can also be painful – it can be the suffering of desire, the pain of rejection, the longing or wanting for mystery and romance. The sensual fills cracks and fissures in the bed-rock of mankind. The sensual is there to keep the body of mankind threaded and alive, and to keep the soul's imagination wakeful. It has no value judgement attached to its energy, for like desire, like love, it is a quality which comes to us when it chooses, not when we choose it.

But are we in danger of losing touch with the sensual at the expense of the material or the purely rational? Have we civilized out the natural senses so that physical and emotional love are empty of soul? It is dangerous to lose sensuality to sexual exploitation, but it is equally dangerous to assume sexuality and sensuality are interchangeable. We may fear others think us effeminate or weak, or over-sexed, or that we are self-indulgent, too bacchanalian, or carnal and wanton. The word sensual is sugared with archaic meanings and decadent associations from which it is impossible to disconnect. In Shakespeare's *Othello*, Iago tried to persuade Rodrigo that the lure of passionate love is controllable by reason: 'if the balance of our lives had not one scale of reason to poise another of sensuality, the blood and baseness of our natures would conduct us to most preposterous conclusions. But we have reason to cool our raging motions, our carnal stings, our unbitted lusts'. Reason may be the counterpoint to lust as Iago suggests, but both are sensual qualities, they are not disparate. Without a 'sense' of consciousness, we would all be animals, bestially anarchic, instinctively primitive and indiscriminate in our carnality. But then

Iago's struggle with his own 'green-eyed monster' was the most demonic of sensual curses, jealousy waxed by shadowed passion.

Even now, even after this book has journeyed through time and ideas to explore the illusion, it is still difficult to break down the notion that 'sensual is woman is carnal is sexual'. So what does this tell us about who we are? If we are so set in our assumptions, are we doomed always to be brain-washed, or is the notion of social indoctrination or presumed conditioning merely an illusion too? Is it in fact perhaps the right way for things to be, that we are seeing sensuality as sexuality because 'incognita', the gods and the *anima mundi* may have conspired for us to think this way?

Doesn't it mean that maybe we do have to balance our senses, that some are reasonable, and some are not? That the forces of demons and gods can chose to enter and overcome us through the very fusion of body and soul, our senses, rather than through the abstraction of the mind? Wouldn't it be so if Vishnu dreams us into being, or the soul sets 'lures and traps' and finds us wanting, that we should have no choice but to be seduced by the pleasures of the flesh rather than by duty, or our conscience?

Doesn't the sensual have a deviousness, a mercurial charm to steal us away from the course of apparent logic, to divert us, to awaken us to desire, to minding the gap that we can never fill? Doesn't it remove us from our subjectivity for a while, to render the ego senselessly sensual? But equally if we were to let ourselves be totally mastered by instinct wouldn't we fall into the Dionysian ecstasy of eternal addiction, chaos, debauchery and madness? Wouldn't the sensual then become no hiding place, no joy, only wretched desire after desire? Eternal pleasure is as terrifying as eternal pain. The balance of being is to honour the separate and yet to see the whole; to have an awareness of ego sense and subjectivity yet to honour the mysteriousness of being. It is to have a true 'sense' of acceptance of self; that which I deem to be my enemy is in fact myself, that which I deem to be my ideal lover is in fact me. It is to hold out one's hand to the mirror and touch, kiss and smell the fragrance of oneself, but to know the sense of oneself is the sense of the Other too.

The sensual quality is innate in us all. As if reflected in a blue glass, it shimmers when held to the light, refracts rainbows when held under water. Consciousness means we can sense the sensual as the underlying force that enables us to relate to the world. Sensuality is in danger of being labelled as merely a decadent art, attached to the world of sexual excess and egotism. But the sensual is androgynous; it speaks a language with no words except those we wish to hear; it has no intention except those which we deem our own. To have total clarity of objective perception is impossible, but the sensual invites us to reflect upon the eternal collusion of dynamic opposites that we have created in our minds, for it is only in our thinking that the 'sense' of opposites lies. The sensual entices us to bring ourselves out of duality through intuition, to see that what I believe to be a stranger is in fact myself, to have a sense of life and death as one, a sense of love and hate as one, a sense of body and mind as one, and always then to have a sense of soul.

NOTE

1. Fortune, pp.59-60

Bibliography

Anderson, Bonnie S., and Judith P. Zinsser, *A History of Their Own*, (2 vols), Penguin books, Harmondsworth, 1990

Ashe, Geoffrey, *The Virgin*, Routledge Kegan & Paul, London, 1976

Baring, Anne, and Jules Cashford, *The Myth of the Goddess*, Arkana Penguin, Harmondsworth, 1993

Briffault, R., *The Mothers: A Study of the Origins of Sentiments and Institutions*, abr. G.R. Taylor, Allen and Unwin, London, 1959

Buonaventura, Wendy, *Belly Dancing*, Virago Press, London, 1983

Budge, Sir E. A., *Wallis, Amulets and Talismans*, N.Y. University Books Inc, New York, 1968

— *Gods of the Egyptians*, (2 vols), Dover Publications, New York, 1969

Campbell, Joseph, *Historical Atlas of World Mythology: Vol 1, The Way of the Animal Powers*, Times Books, London, 1984

Chagnon, N. A., *Yanomamo: The Fierce People*, Holt, Rinehart & Winston, New York, 1968

Daly, Mary, *Beyond God the Father*, Beacon Press, Boston, 1973

de Voragine, Jacobus, *The Golden Legend*, Longmans, Green & Co, New York, 1941

Dijkstra, Bram, *Idols of Perversity*, Oxford University Press, New York, 1986

Edwards, Owen Dudley (ed.), *The Fireworks of Oscar Wilde*, Barrie & Jenkins Ltd, London, 1991

Edwardes, A., *The Jewel in the Lotus*, Lancer Books, New York, 1965

Euripedes, trans. Philip Vellacott, Alcestis/Hippolytus, Penguin Classics, Harmondsworth, 1974

Fantham, Elaine, Foley, et al, *Women in the Classical World*, Oxford University Press, New York, 1994

Flaubert, Gustave, *Madame Bovary*, trans. Gerard Hopkins, Oxford University Press, Oxford, 1981

Flaubert, Gustave, *Herodias*, trans. R. Baldick, Penguin, Harmondsworth, 1961

Flaubert, Gustave, *Salammbo*, trans. Krailsheimer, Penguin, Harmondsworth, 1977

Fortune, Dion, *Moon Magic*, Samuel Weiser, York Beach, 1956

Friedrich, Paul, *The Meaning of Aphrodite*, University of Chicago Press, Chicago, 1978

Gadon, E. W., *The Once and Future Goddess: A Symbol For Our Time*, Aquarian Press, Wellingborough, 1990

Goodison, Lucy, and Christine Morris, eds., *Ancient Goddesses*, British Museum Press, London, 1998

Hillman, James, *The Thought of the Heart*, Spring Publications Inc., Dallas, 1981

— *Revisioning Psychology*, Harper and Row, New York, 1976

Johnson, Buffie, *Lady of the Beasts: Ancient Images of the Goddess and Her Sacred Animals*, Harper and Row, San Francisco, 1988

Jung, Carl, *Collected Works*, 20 vols, eds. Sir Herbert Read et al., trans. R. F. C. Hull, Routledge & Kegan Paul, London, 1957–79

— *Aspects of the Feminine (1982)*, Routledge, London 1986.

Koltuv, Barbara Black, *The Book of Lilith*, Nicolas-Hays Inc, York Beach, 1986

Kramer, S., *The Sacred Marriage*, Indiana University Press, Indiana, 1969

Lawrence, D. H, *The Rainbow*, Penguin Books, Harmondsworth, 1971.

Lefkowitz, Mary R., and Maureen B. Fant, *Women's Life in Greece and Rome*, The Johns Hopkins University Press, Baltimore, 1982

Lindsay, J., *The Ancient World*, G. P. Putnam's & Sons, New York, 1968

Lustbader, Eric Van, *The Ninja*, Granada Publishing Ltd., St Albans, 1981

BIBLIOGRAPHY

McCulloch, Colleen, *The Thorn Birds*, Futura Publications Ltd., London, 1978

Mann, A. T., and Jane Lyle, *Sacred Sexuality*, Element Books, Shaftesbury, 1995

Marshack, A., *The Roots of Civilisation*, Weidenfeld & Nicholson, London, 1972

Moore, T. (ed.), *The Essential James Hillman*, Routledge, London, 1990

Paglia, Camille, *Sexual Personae*, Penguin Books, Harmondsworth, 1992

Praz, Mario, *The Romantic Agony*, London 1933, new ed., New York, 1951.

Radice, Betty, trans., *The Letters of Abelard and Héloise*, Penguin Classics, Harmondsworth, 1974

Ramsay, Jay, *Alchemy, The Art of Transformation*, Thorsons/Harper Collins, London 1997

Rilke, R. M., *Selected Poetry*, ed. Stephen Marshall, Picador, London, 1982

Roberts, Jane, *The Nature of the Psyche*, Prentice Hall, New York, 1979

Robinson, James M. (ed.), *Nag Hammadi Library in English* (2nd ed.), trans. Members of the Coptic Gnostic Library Project of the Institute for Antiquity and Christianity, E. J. Brill, Leiden, 1984

Sacher-Masoch, Leopold von, *Venus in Furs* (1870), trans. John Glassco, Blackfish Press, Burnaby, 1970

Silbever, Herbert, *Hidden Symbolism of Alchemy and the Occult Arts*, Dover Publications, New York, 1971

Stoker, Bram, *Dracula*, Signet Classics, New York, 1965

Symons, Arthur, *Selected Poetry and Prose*, Intro. R. V. Holdsworth, Carcanet Press, Cheadle Thelme, 1974

Tannahill, Reay, *Sex in History*, Book Club Associates, London, 1980

Thornton, Bruce S., *Eros, The Myth of Ancient Greek Sexuality*, Westview Press, Colorado, 1997

Walker, Barbara G., *The Women's Encyclopedia of Myths and Secrets*, Pandora, London, 1983

Warner, Marina, *Alone of All Her Sex: The Myth and Cult of the Virgin Mary*, Picador (Pan Books), London, 1985

Watterson, Barbara, *Women in Ancient Egypt*, Sutton Publishing, Stroud, 1994

Watts, Alan, *Myth and Ritual in Christianity*, Thames and Hudson, London, 1954

Wender, Dorothea, trans., *Hesiod and Theognis*, Penguin, Harmondsworth, 1973

Wilber, Ken, *No Boundary*, Shambhala Publications, Boston, 1979

Wright, M. R., *Cosmology in Antiquity*, Routledge, London, 1995

Young, Serenity, *An Anthology of Sacred Texts By and About Women*, Pandora/HarperCollins, London 1993

Yu, Li, *The Before Midnight Scholar*, trans. Richard Martin, Arrow Books, London, 1965

Zola, Emile, *Nana*, trans. George Holden, Penguin Books, Harmondsworth, 1972

Index

A-sensual, 143–50
Abelard, and Hèloise, 131–5
Aborgine, creation myth, 20
Adornment, 71–82
    Amazonian, 76
    Aztec, 75
    erotic, 73, 85
    Etruscan, 77
    Egyptian, 75
    and feelers, 72
    purpose, 81
    Renaissance, 78
    sexual, 83
    worldwide, 74–82
Androgyny, 251, 255, 260
Aphrodite, 83–5, 270, 88–9, 235, 269, 270
Apollo, 93, 188
Archetypes, 8, 15, 23, 220
Atalanta, 269
Attraction, sexual, 83–89

Bardot, Brigitte, 233
Beauty, 53–69
    and Age, 59
    and the Beholder, 56
    Covered up, 57
    Chinese, 62
    and eternal youth, 66
    harmony, 61
    inner, 77
    Indian, 62
    value of, 68–9

Belly-dancing, 36
Betrayal, 202–3
Blame, Who?, 124–5
Blood, 39
Body, human 13–29 *passim*
    man's, 24–7, 193,
    woman's, 23–6, 235
    sense, 15
Borgia, Lucretia, 238, 269
Bovary, Emma, 239–42
Braun, Eva, 238
Breasts, 148–9
Byron, Lord, 244, 263

Calypso, 273
Casanova, 244
Charisma, 250–2
Christianity, 115–16
    and fear of women, 126, 132, 138, 140
    and Eve, 115–16, 118–19
    and Mary, 143–8, 149, 150, 154
    and repression of women, 146–8, 150, 154, 159–64
Circe, 268
Civilisation, 58
Cleopatra, 220–2, 256
Clytemnestra, 97, 232, 248
Colet, Louise, 273
Corday, Charlotte, 238
Courtesans, 105–6
Courtly love, 171, 178–83

Danger, meaning of, 239
  sensuality in, 277, 278
Desire, 95, 109–11, 134–6, 186–91,
  194, 195, 197, 209, 275
Dionysus, 93, 156, 165, 274, 275
Disenchantment, 97
Duncan, Isadora, 258–9, 269

East, as sensual, 262–3
Echo, as personfication, 253–4
Eco-relating, 129–31
'Egoic' sense, 16
Egyptian women, 75, 80
Epithemia, 183–4, 270
Embodiment, of the feminine
  archetype, 28, 57, 82, 102, 197,
  202, 231–2, 235, 262
Eros, 46, 84, 105, 190, 197–8, 204–5,
  212
Eroticism and sensuality, 86, 94,
  176, 190–1, 192–5, 205–6
Eurynome, 32
Eve, 73, 115–16, 118–19

Fate, 59, 194, 231–42, 266
  as woman, 137–40, 194
Female Initiator, 31
Femaleness, 58, 86, 128
Feminine, archetype, 28, 57, 82,
  102, 197, 202, 231–2, 235,
  262
Feminine, the dark, 113–25, 152,
  198, 223, 231, 235, 246
Femme fatale, 117, 152, 194,
  231–48
Fig leaf, 72
Flagellation, 242–5
Flaubert, 239–40, 273

Garbo, Greta, 259
Gender, 7, 13, 17, 18, 19, 152, 199,
  255, 277
Geishas, 106–7
Gilgamesh, 66
Glamour, 249–64
Goddess, 24, 34
Guilhelm, 179–80

Hair, as sensual, 259–61
Harriman, Pamela, 246
Hedonism, 266
Helen of Troy, 90–2
Heloise, 131–5
Heraclitus, 20
Heresy, 160–4
Hesiod, 102
Hope, as eroticising mankind,
  104–5

Inanna, 34, 35
Inhibition, sensual, 165–6
Inquisition, 159–64
Insecurity with the body, 37–8
Intuition as a feminine quality, 15
Incognita as a sensual quality, 128,
  131, 263

Joan of Arc, 162–3
Jordan, Queen Noor of, 262
Jung, 23, 33, 44, 46

Kahn, Jemima, 262
Kissing, 206–11

Lamb, Lady Caroline, 263
Lilith, 115–17, 124–25, 231, 234, 272
Loneliness, 196–7

Love, courtly, 171, 178–83
  earthly and sensual, 164, 171, 178–84, 189–225
  and Mystery, 127–131, 214–8
  desire, 94
  Romantic, 164,175–91, 195

Makropoulos Secret, The, 189–90
Male, sensuality, 59, 153, 241, 170, 226–7
Man as fate, 244–5
Mansfield, Jane, 233
Manson, Charles, 257
Marduk, 32
Marriage, 97, 182, 276
Mary the Virgin, 143–5
Mary Magdalene, 150–2
Mata Hari, 238
Maya, 212, 213
Medusa, 49
Menstruation, 39–43
Minoans, and adornment, 57, 74
Misogyny, 92, 97, 101, 145
Mother, the trouble with, 44–7
  or lover, 166–8
  and wooing, 169, 170
Mother Teresa, 136
Mystics, female, 135
Mystery, woman as personfication of, 17, 127–31, 205, 214–18
Myth, 32–6

Nana, 209
Narcissus, as personification, 253–4
Nature, split from, 21
Neolithic peoples, 21
Nin, Anais, 269

Ninja, The, 138
Nyx, 261

Ophion, 32

Pain, woman as personification of, 242–5
Pandora, 103–5
Pasiphae, 270
Passion, sensual, 200, 203, 223
Phaedra, 199–200
  New, 200–2, 203–4
Phallus, 27
Pheromones, 80, 93
Phryne, 237–8
Plato, 22, 65
Plutarch, 64
Pompadour, Madame de, 258
Proteus, 256

Relationship, sensuality and, 175– 225
Renaissance, 78, 202, 267
Romance, 171, 177–8, 180–9
Reine Margot, 234

Salome, 154–5, 232
Sappho, 88
Seduction, 213–227 *passim*
Sense, meaning, 14
  of disenchantment, 96
Senses, five, 14, 15
Sensual awareness, 87, 265–78
  Charisma, 250–2
  desire, 109–11
  Glamour, 249–56
  'hunting', 274–5
  love, 17, 175–225

INDEX

mystery, 17, 127–31, 205, 216, 218, 252
needs, 273–4
power, 135, 257–8
seduction, 213–27
sexuality and, 83–9, 101–4, 135, 150, 161, 175–6, 211, 241, 266–7, 276–7
suppression, 159–71
'types', 266–73
value, 175–91, 192–227, 265–80
Sensuality
and men, 59, 153, 170, 226–7, 241
woman as personification of, 43, 101, 235
women as stereotypes of, 26, 152, 199, 231, 267, 277
Serpent, 47, 85
Shiva, 93
Sirens, 216–18, 223
Shadows, psychological, 114
Church, 115
Slinky, woman as, 48–9
Snakes, 47, 85
Sphinx, of Thebes, 117
Symbols, 13, 22

Taoism and sensuality, 20, 60, 62, 95, 211–12
Thargelia, 107–8

Theodora, 108–9
Tiamut, 31
Transitional Object, psychological, 89–90
Tristan and Iseult, 186–9
Thorn Birds, The, 139
Throttler, the, 236–7

Vampires, 119–22
Virgin, Whore dilemma, 143–57, 175
Virgin Mary, 143–48, 149, 150, 154

Womb, 25
Woman, as fate, 137–40, 194, 231–42
as mystery, 17, 127–31, 205, 214–18
as symbol of the feminine, 28, 57, 82, 102, 197, 198, 199, 202, 215, 216, 223, 235, 277
Women, repression of, 146–8, 150, 154, 175
Whore, Virgin dilemma, 143–57, 175
Witch-hunts, 159–64, 170–1

Yin/Yang, 19, 28, 255, 277

Zeus, 156, 275